TENSOR ANALYSIS

BY

EDWARD NELSON

PRINCETON UNIVERSITY PRESS

AND THE

UNIVERSITY OF TOKYO PRESS

PRINCETON, NEW JERSEY

1967

Published in Japan exclusively by the
University of Tokyo Press;
in other parts of the world by
Princeton University Press

Printed in the United States of America

Preface

These are the lecture notes for the first part of a one-term course on differential geometry given at Princeton in the spring of 1967. They are an expository account of the formal algebraic aspects of tensor analysis using both modern and classical notations.

I gave the course primarily to teach myself. One difficulty in learning differential geometry (as well as the source of its great beauty) is the interplay of algebra, geometry, and analysis. In the first part of the course I presented the algebraic aspects of the study of the most familiar kinds of structure on a differentiable manifold and in the second part of the course (not covered by these notes) discussed some of the geometric and analytic techniques.

These notes may be useful to other beginners in conjunction with a book on differential geometry, such as that of Helgason [2,§1], Nomizu [5,§5], De Rham [7,§7], Sternberg [9,§8], or Lichnerowicz [11,§9]. These books, together with the beautiful survey by S. S. Chern of the topics of current interest in differential geometry (Bull. Am. Math. Soc., vol. 72, pp. 167-219, 1966) were the main sources for the course.

The principal object of interest in tensor analysis is the module of C^∞ contravariant vector fields on a C^∞ manifold over the algebra of C^∞ real functions on the manifold, the module being equipped with the additional structure of the Lie product. The fact that this module is "totally reflexive" (i.e. that multilinear functionals on it and its dual can be identified with elements of tensor product modules) follows-for a finite-dimensional second-countable

C^∞ Hausdorff manifold - by the theorem that such a manifold has a covering by finitely many coordinate neighborhoods. See J. R. Munkres, Elementary Differential Topology, p.18, Annals of Mathematics Studies No. 54, Princeton University Press, 1963.

I wish to thank the members of the class, particularly Barry Simon, for many improvements, and Elizabeth Epstein for typing the manuscript so beautifully.

CONTENTS

§1. Multilinear algebra

1. The algebra of scalars

We make the permanent conventions that F^0 is a field of characteristic 0 and that F is a commutative algebra with identity over F^0. Elements of F will be called <u>scalars</u> and elements of F^0 will be called <u>constants</u>.

The main example we have in mind is F^0 the field $\mathbb{R}$ of real numbers and F the algebra of all real C^∞ functions on a C^∞ manifold M. In this example the set of all C^∞ contravariant vector fields is a module over F, with the additional structure that the contravariant vector fields act on the scalars via differentiation and on each other via the Lie product. Tensor analysis is the study of this structure. In this section we will consider only the module structure.

2. Modules

The term "module" will always mean a unitary module $(1X = X)$. Thus an F module E is an Abelian group (written additively) with a mapping of $F \times E$ into E (indicated by juxtaposition) such that

$$f(X+Y) = fX+fY ,$$

$$(f+g)X = fX+gX ,$$

$$(fg)X = f(gX) ,$$

$$1X = X ,$$

for all X,Y in E and f,g in F.

If E is an F module, the <u>dual module</u> E' is the module of all

F-linear mappings of E into F. If $\omega \in E'$ we denote the value of ω on X in E by any of the symbols

$$\omega(X)\ , \quad X(\omega)\ , \quad <\omega,X>\ , \quad <X,\omega>\ .$$

If A is an F-linear mapping of E into E its <u>dual</u> A' defined by

$$<A'\omega,X> = <\omega,AX>$$

is an F-linear mapping of E' into E'. There is a natural mapping $\kappa: E \longrightarrow E''$ defined by

$$(\kappa X)(\omega) = <\omega,X>\ , \qquad \omega \in E'\ ,$$

and E is called <u>reflexive</u> in case κ is bijective. (The mapping κ is not in general injective. For example, if F is the algebra of all C^∞ functions on a manifold M and E is the F module of all continuous contravariant vector fields then $E' = 0$.)

The notions of <u>submodule</u>, F <u>module homomorphism</u>, and <u>quotient module</u> are defined in the obvious way. If H and K are F modules and $\pi: H \longrightarrow K$ is an F module homomorphism then the quotient module $H/\ker \pi$ is canonically isomorphic to the image of π. See Bourbaki [1].

We will frequently refer to the elements X of an F module E as <u>contravariant vector fields</u> or <u>vector fields</u> and to elements ω of the dual module E' as <u>covariant vector fields</u> or <u>1-forms</u>.

3. Tensor products

If H and K are two F modules, their tensor product $H \otimes K$ (over F) is the F module whose Abelian group is the free Abelian group generated by all pairs $X \otimes Y$ with X in H and Y in K modulo the subgroup generated by all elements of the form

$$
(X_1+X_2)\otimes Y - X_1\otimes Y - X_2\otimes Y\ ,
$$
(1)
$$
X\otimes(Y_1+Y_2) - X\otimes Y_1 - X\otimes Y_2\ ,
$$
$$
(fX)\otimes Y - X\otimes(fY)
$$

where f is in F, and the action of F on $H\otimes K$ is given by

$$
f(X\otimes Y) = (fX)\otimes Y = X\otimes(fY)\ .
$$

Let E be an F module. We define

$$
\mathring{E}_r^s = E\otimes\ldots\otimes E\otimes E'\otimes\ldots\otimes E' \qquad (E\ r\ \text{times},\ E'\ s\ \text{times}).
$$

If r or s is 0 we sometimes omit it, and we set $\mathring{E}_0^0 = \mathring{E}^0 = \mathring{E}_0 = F$. Notice that $\mathring{E}_1 = E$, $\mathring{E}^1 = E'$. We also define

$$
\mathring{E}_* = \sum_{r=0}^{\infty} \mathring{E}_r\ ,
$$
$$
\mathring{E}^* = \sum_{r=0}^{\infty} \mathring{E}^r\ ,
$$
$$
\mathring{E}_*^* = \sum_{r,s=0}^{\infty} \mathring{E}_r^s\ ,
$$

where the sums are weak direct sums (only finitely many components of any element are non-zero).

Notice that $\mathring{E}_*$ and $\mathring{E}^*$ are associative graded F algebras with the tensor product $\otimes$ as multiplication. We make the identification

$$
\mathring{E}_r\otimes\mathring{E}^s = \mathring{E}^s\otimes\mathring{E}_r\ .
$$

With this identification, $\mathring{E}_*^*$ is an associative bi-graded F-algebra.

4. Multilinear functionals

Let E be an F-module. We define E_r^s to be the set of all F-multilinear mappings of

$$
E'\times\ldots\times E'\times E\times\ldots\times E \qquad (E'\ r\ \text{times},\ E\ s\ \text{times})
$$

into F. Thus if $u \in E_r^s$,

$$u(\omega^1,\ldots,\omega^r,X_1,\ldots,X_s), \qquad \omega^i \in E', \quad X_j \in E,$$

is a scalar, and if all arguments but one are held fixed its value depends in an F-linear way on the remaining argument. With the obvious definitions of addition and scalar multiplication, E_r^s is an F-module. If r or s is 0 we sometimes omit it, and we set $E_0^0 = E^0 = E_0 = F$. Notice that $E^1 = E'$ and $E_1 = E''$. We also define the weak direct sums

$$E_* = \sum_{r=0}^{\infty} E_r,$$

$$E^* = \sum_{r=0}^{\infty} E^r$$

$$E_*^* = \sum_{r,s=0}^{\infty} E_r^s .$$

For u in E_r^s, v in $E_{r'}^{s'}$ we define $u \otimes v$ (this is a different use of the symbol $\otimes$) in $E_{r+r'}^{s+s'}$ by

$$(u\otimes v)(\omega^1,\ldots,\omega^{r+r'},X_1,\ldots,X_{s+s'})$$
$$= u(\omega^1,\ldots,\omega^r,X_1,\ldots,X_s)v(\omega^{r+1},\ldots,\omega^{r+r'},X_{s+1},\ldots,X_{s+s'}) .$$

Then E_* and E^* are associative graded F algebras and E_*^* is an associative bi-graded F algebra, all with $\otimes$ as multiplication.

5. Two notions of tensor field

The preceding paragraphs suggest two different notions of a tensor field: an element of $\overset{\circ *}{E}_*$ or an element of E_*^*. Happily, the two notions coincide for finite dimensional differentiable manifolds (assumed to be paracompact). The second notion is of greater importance, so that if E is an F module we will refer to elements of E_*^* as <u>tensor fields</u> or <u>tensors</u>.

A tensor in E_r^s is said to be <u>contravariant of rank</u> r and <u>covariant of rank</u> s, E_* is the <u>contravariant tensor algebra</u>, E^* the <u>covariant tensor algebra</u>, and E_*^* the <u>mixed tensor algebra</u>.

There is a natural F algebra homomorphism

$$\kappa\colon \overset{\circ}{E}{}_*^* \longrightarrow E_*^*,$$

preserving the bi-grading, defined by setting

$$(\kappa(X_1\otimes\ldots\otimes X_r\otimes\omega^1\otimes\ldots\otimes\omega^s))(\eta^1,\ldots,\eta^r,Y_1,\ldots,Y_s)$$
$$= X_1(\eta^1)\ldots X_r(\eta^r)\omega^1(Y_1)\ldots\omega^s(Y_s)$$

and extending to all of $\overset{\circ}{E}{}_*^*$ by F linearity. By the definition of tensor product, κ is well-defined. This agrees with our previous definition of κ as a mapping of $E = \overset{\circ}{E}_1$ into $E'' = E_1$. We call the F module E <u>totally reflexive</u> in case κ is bijective. As mentioned before, the module of all C^∞ contravariant vector fields on a finite dimensional paracompact manifold is totally reflexive.

6. F-linear mappings of tensors

<u>Theorem 1. Let</u> E <u>be an</u> F <u>module. Then</u> $E_{s+r'}^{r+s'}$ <u>is canonically isomorphic to the</u> F <u>module</u>

$$\mathrm{Hom}_F(\overset{\circ}{E}{}_r^s,\ E_{r'}^{s'})$$

<u>of all F-linear mappings of</u> $\overset{\circ}{E}{}_r^s$ <u>into</u> $E_{r'}^{s'}$. <u>The isomorphism</u> $\iota = \iota(r,s,r',s')$ <u>is defined by setting</u>

$$(2)\qquad ((\iota u)(X_1\otimes\ldots\otimes X_r\otimes\omega^1\otimes\ldots\otimes\omega^s))(\eta^1,\ldots,\eta^{r'},Y_1,\ldots,Y_{s'})$$
$$= u(\omega^1,\ldots,\omega^s,\eta^1,\ldots,\eta^{r'},X_1,\ldots,X_r,Y_1,\ldots,Y_{s'})$$

<u>and extending to all of $\overset{\circ}{E}{}^s_r$ by F-linearity. In particular, the dual module of $\overset{\circ}{E}{}^s_r$ is canonically isomorphic to E^r_s, so that if E is totally reflexive then each E^s_r is reflexive.</u>

<u>Proof.</u> The mapping ι is well-defined by the definition of tensor product, and is an F module homomorphism. It is obviously injective and surjective. QED.

Suppose that E is totally reflexive. A number of special cases of Theorem 1 come up sufficiently often to warrant discussion. We identify $(E^s_r)'$ and E^r_s, and denote the pairing by any of the expressions $< u,v >$, $< v,u >$, $u(v)$, $v(u)$, as convenient. If A is in E^1_1 we use the same symbol A for the F-linear transformation $\iota(1,0,1,0)A$ of $E = E_1$ into itself, so that

$$< \omega, AX > = A(\omega, X) , \qquad A \in E^1_1 .$$

Notice that the F-linear transformation $\iota(0,1,0,1)A$ of E^1 into E^1 is the dual A' of A. If A and B are in E^1_1 we write AB for their product as F-linear mappings of E into itself and similarly for A^n. The identity mapping of E into itself is denoted 1.

If E is totally reflexive then $\iota(2,0,1,0)$ identifies E^2_1 with the set of all structures of F algebra (not necessarily associative) on E. If B is in E^2_1 we write $B(X,Y)$ or $B_X Y$ for the product in this sense of the two vector fields X and Y, so that

$$< \omega, B(X,Y)> = < \omega, B_X Y > = B(\omega, X, Y) , \qquad B \in E^2_1 .$$

Also, $\iota(1,0,0,1)$ identifies E^2 with the set of all F-linear mappings of E_1 into E^1, so that

$$< u(X), Y > = u(X,Y) , \qquad u \in E^2 .$$

Similarly, $\iota(0,1,1,0)$ identifies E_2 with the set of all F-linear mappings of E^1 into E_1, so that

$$< \eta, v(\omega)> = v(\omega,\eta) , \qquad v \in E_2 .$$

7. Contractions

Let E be an F module, and let $1 \leq \mu \leq r$, $1 \leq \nu \leq s$. We define the <u>contraction</u>

$$C^\mu_\nu : \mathring{E}^s_r \longrightarrow \mathring{E}^{s-1}_{r-1}$$

by

$$C^\mu_\nu(X_1 \otimes \ldots \otimes X_r \otimes \omega^1 \otimes \ldots \otimes \omega^s)$$

$$= X_\mu(\omega^\nu) X_1 \otimes \ldots \otimes \hat{X}_\mu \otimes \ldots \otimes X_r \otimes \omega^1 \otimes \ldots \otimes \hat{\omega}^\nu \otimes \ldots \otimes \omega^s ,$$

where the circumflex denotes omission, and by extending C^μ_ν to all of $\mathring{E}^s_r$ by F-linearity. By the definition of tensor product, this is well-defined, and it is a module homomorphism. The Encyclopaedia Britannica calls it an operation of almost magical efficiency. (See the interesting article on tensor analysis in the 14th edition.)

If $A \in E^1_1$ then $C^1_1 A$ is denoted $\operatorname{tr} A$, and called the <u>trace</u> of A.

8. The symmetric tensor algebra

Let E be an F module and let $\mathfrak{S}(r)$ be the symmetric group on r letters. For u in E_r and σ in $\mathfrak{S}(r)$ define $u\rho(\sigma)$ by

$$(u\rho(\sigma))(\omega^1,\ldots,\omega^r) = u(\omega^{\sigma(1)},\ldots,u^{\sigma(r)}), \qquad \omega^i \in E^1 .$$

Then ρ is a right representation of $\mathfrak{S}(r)$ on E_r ; that is,

(3) $$u\rho(\sigma\tau) = u\rho(\sigma)\rho(\tau) .$$

Define Sym on E_r by

$$\text{Sym } u = \frac{1}{r!} \sum_{\sigma} u\rho(\sigma) .$$

(Since F^0 is a field of characteristic zero, $1/r!$ makes sense.) Extend Sym to the contravariant tensor algebra E_* by additivity. A contravariant tensor u is called <u>symmetric</u> in case $\text{Sym } u = u$. Thus u in E_r is symmetric if and only if $u(\omega^1,\ldots,\omega^r)$ is invariant under the transposition of any pair of ω's. The set of all symmetric tensors in E_r is denoted S_r and the set of all symmetric tensors in E_* is denoted S_*, so that

$$S_* = \sum_{r=0}^{\infty} S_r$$

where of course $S_0 = F$.

<u>Theorem 2.</u> <u>Sym is F-linear and is a projection</u> $(\text{Sym}^2 = \text{Sym})$ <u>with range</u> S_*. <u>Consequently</u> S_* <u>may be identified with the quotient of</u> E_* <u>by the kernel of</u> Sym. <u>The kernel of</u> Sym <u>is a two-sided ideal in</u> E_*. <u>Consequently the multiplication</u>

$$uv = \text{Sym } u\otimes v \tag{4}$$

<u>makes</u> S_* <u>into an associative commutative graded algebra over</u> F.

<u>Proof.</u> Sym is clearly F-linear. That it is a projection follows from (3) - it is easily checked that the average over a group representation is a projection. The range of Sym is S_* by definition, so that we may identify S_* with the quotient of E_* by the kernel of Sym.

By the definitions of $\otimes$ and Sym, if $u \in E_r$ and $v \in E_s$ then

$$(\text{Sym } u\otimes v)(\omega^1,\ldots,\omega^{r+s}) = \frac{1}{(r+s)!} \sum_{\sigma} u(\omega^{\sigma(1)},\ldots,\omega^{\sigma(r)})v(\omega^{\sigma(r+1)},\ldots,\omega^{\sigma(r+s)}) \tag{5}$$

where σ ranges over $\mathfrak{S}(r+s)$. If Sym u or Sym v is 0 this is clearly

0 , so that the kernel of Sym is a two-sided ideal, and the quotient algebra is an associative commutative graded F algebra. QED.

The algebra S_* is called the (contravariant) <u>symmetric tensor algebra.</u> One may also construct the covariant symmetric tensor algebra S^*.

9. The Grassmann algebra

The discussion of the (covariant) Grassmann algebra, given an F module E , proceeds along similar lines. For α in E^r and σ in $\mathfrak{S}(r)$, define $\alpha\tilde{\rho}(\sigma)$ by

$$(\alpha\tilde{\rho}(\sigma))(X_1,\ldots,X_r) = (\operatorname{sgn}\sigma)\alpha(X_{\sigma(1)},\ldots,X_{\sigma(r)}) , \qquad X_i \in E$$

where $\operatorname{sgn}\sigma$ is 1 for σ an even permutation and -1 for σ an odd permutation. Then $\tilde{\rho}$ is a right representation of $\mathfrak{S}(r)$ on E^r . Define Alt on E^r by

$$\operatorname{Alt}\alpha = \frac{1}{r!}\sum_{\sigma}\alpha\tilde{\rho}(\sigma)$$

and extend Alt by additivity to E^* . An element α of E^* such that $\operatorname{Alt}\alpha = \alpha$ is called <u>alternate</u> or <u>antisymmetric</u> and is also called an <u>exterior form.</u> The set of alternate tensors in E^r is denoted A^r , and elements of A^r are called <u>r-forms.</u> The set of all alternate tensors in E^* is denoted A^* , so that

$$A^* = \sum_{r=0}^{\infty} A^r .$$

Notice that $A^0 = F$ and $A^1 = E^1$. A covariant tensor α of rank r is alternate if and only if $\alpha(X_1,\ldots,X_r)$ changes sign under the transposition of any two X's .

<u>Theorem 3.</u> Alt <u>is F-linear and is a projection with range</u> A^* . <u>Consequently</u> A^* <u>may be identified with the quotient of</u> E^* <u>by the kernel</u>

of Alt . The kernel of Alt is a two-sided ideal in E^* and the multiplication

$$\alpha\wedge\beta = \text{Alt } \alpha\otimes\beta$$

makes A^* into an associative graded algebra over F satisfying

$$(6) \qquad \beta\wedge\alpha = (-1)^{rs}\, \alpha\wedge\beta\,, \qquad \alpha \in A^r, \quad \beta \in A^s\,.$$

Proof. The proof is quite analogous to the proof of Theorem 2. Instead of (5) we have, for α in E^r and β in E^s,

$$(7) \qquad (\text{Alt } \alpha\otimes\beta)(X_1,\ldots,X_{r+s})$$
$$= \frac{1}{(r+s)!} \sum_{\sigma} (\text{sgn } \sigma)\alpha(X_{\sigma(1)},\ldots,X_{\sigma(r)})\beta(X_{\sigma(r+1)},\ldots,X_{\sigma(r+s)})\,.$$

QED.

The algebra A^* is called the (covariant) Grassmann algebra. One can also construct the contravariant Grassmann algebra A_* .

Warning: As we have defined the notion, an r-form is simply a covariant tensor of rank r which is alternate. However it is customary in the literature, and we will follow the custom because it is convenient, to make from time to time conventions about r-forms which differ from conventions already made about tensors. These special conventions have the purpose of ridding the notation of factors $r!$, etc.

If α is an exterior form we denote by α^k the exterior product of α with itself k times, $\alpha^k = \alpha\wedge\ldots\wedge\alpha$. If $k > 1$ this is 0 for α in A^1 or for α an exterior product of 1-forms, but not for general elements of A^r . Notice that $f\wedge\alpha = \alpha\wedge f = f\alpha$ for f in $F = A^0$ and α in A^* .

A graded algebra whose multiplication satisfies (6) is sometimes called "commutative," but this miserable terminology will not be used here.

10. Interior multiplication

Let X be in the F module E and let α be an r-form. We define $X \lrcorner \alpha$ by

(8) $$(X \lrcorner \alpha)(X_2,\ldots,X_r) = r\alpha(X,X_2,\ldots,X_r) ,$$

$X \lrcorner \alpha = 0$ if $\alpha \in A^0$, and we define $X \lrcorner \alpha$ by additivity if α is a general element of A^*. The mapping $\alpha \longrightarrow X \lrcorner \alpha$ is F-linear from A^r to A^{r-1}, and it follows from (7) that it is an antiderivation of A^*; that is,

(9) $$X \lrcorner (\alpha \wedge \beta) = (X \lrcorner \alpha) \wedge \beta + (-1)^r \alpha \wedge (X \lrcorner \beta) , \qquad \alpha \in A^r, \quad \beta \in A^* .$$

11. Free modules of finite type

An F module E is <u>free of finite type</u> if there exist $X_1,\ldots,X_n$ in E, called a <u>basis</u>, such that every element Y in E has a unique expression of the form

$$Y = \Sigma\, Y^i X_i , \qquad Y^i \in F .$$

(Unless indicated otherwise, Σ always denotes summation over all repeated indices.)

<u>Theorem 4.</u> <u>Let</u> E <u>be free of finite type, with a basis</u> $X_1,\ldots,X_n$. <u>Then</u> E <u>is totally reflexive. The dual module has a unique basis</u> $\omega^1,\ldots,\omega^n$ <u>(called the dual basis) such that</u>

$$< \omega^i, X_j > = \delta^i_j ,$$

<u>where</u> δ^i_j <u>is</u> 1 <u>if</u> $i = j$ <u>and</u> 0 <u>otherwise. The</u>

$$X_{i_1} \otimes \ldots \otimes X_{i_r} \otimes \omega^{j_1} \otimes \ldots \otimes \omega^{j_s}$$

<u>are a basis of</u> E^s_r, <u>so that every</u> u <u>in</u> E^s_r <u>has a unique expression of</u>

the form

$$u = \Sigma\, u^{i_1 \dots i_n}_{j_1 \dots j_s} X_{i_1} \otimes \dots \otimes X_{i_r} \otimes \omega^{j_1} \otimes \dots \otimes \omega^{j_s} .$$

The coefficients in this expression (called the components of u with respect to the given basis of E) are given by

$$u^{i_1 \dots i_r}_{j_1 \dots j_s} = u(\omega^{i_1}, \dots, \omega^{i_r}, X_{j_1}, \dots, X_{j_s}) .$$

The

(10) $$\omega^{i_1} \wedge \dots \wedge \omega^{i_r} , \qquad i_1 < \dots < i_r$$

are a basis of A^r, so that every r-form α has a unique expression of the form

$$\alpha = \sum_{i_1 < \dots < i_r} \alpha_{i_1 \dots i_r} \omega^{i_1} \wedge \dots \wedge \omega^{i_r} .$$

The coefficients in this expression (called the components of α as an r-form or simply the components of α) are given by

$$\alpha_{i_1 \dots i_r} = r!\, \alpha(X_{i_1}, \dots, X_{i_r}) ,$$

so that the components of α as an r-form are $r!$ times the components of α regarded as an element of E^r. If $r > n$ then $A^r = 0$.

If $u \in E^s_r$, $v \in E^{s'}_{r'}$, then $u \otimes v$ has components

$$(u \otimes v)^{i_1 \dots i_r k_1 \dots k_{r'}}_{j_1 \dots j_s \ell_1 \dots \ell_{s'}} = u^{i_1 \dots i_r}_{j_1 \dots j_s} v^{k_1 \dots k_{r'}}_{\ell_1 \dots \ell_{s'}} .$$

If $1 \le \mu \le r$, $1 \le \nu \le s$ and $u \in E^s_r$ then $C^\mu_\nu u$ has components

$$(C^\mu_\nu u)^{i_1 \dots \hat{i}_\mu \dots i_r}_{j_1 \dots \hat{j}_\nu \dots j_s} = \Sigma\, u^{i_1 \dots i_{\mu-1} a i_{\mu+1} \dots i_r}_{j_1 \dots j_{\nu-1} a j_{\nu+1} \dots j_s} .$$

If $u \in E^{r+s'}_{s+r'}$ and $v \in E^{s}_{r}$ then $(\iota u)v$ in $E^{s'}_{r'}$ has components

$$((\iota u)v)^{k_1 \dots k_{r'}}_{\ell_1 \dots \ell_{s'}} = \Sigma\, u^{i_1 \dots i_s k_1 \dots k_{r'}}_{j_1 \dots j_r \ell_1 \dots \ell_{s'}} v^{j_1 \dots j_r}_{i_1 \dots i_s} .$$

If $u \in E_r$ then Sym u has components

$$(\mathrm{Sym}\ u)^{i_1 \dots i_r} = \frac{1}{r!} \Sigma\, \varepsilon^{i_1 \dots i_r}_{j_1 \dots j_r} u^{j_1 \dots j_r}$$

where the ε is 1 if the j's are a permutation of the i's and is 0 otherwise. If $\alpha \in E^r$ then Alt α has components as an element of E^r given by

$$(\mathrm{Alt}\ \alpha)_{i_1 \dots i_r} = \frac{1}{r!} \Sigma\, \delta^{j_1 \dots j_r}_{i_1 \dots i_r} \alpha_{j_1 \dots j_r}$$

where the δ is 1 if the j's are an even permutation of the i's, is -1 if the j's are an odd permutation of the i's, and is 0 otherwise. If α is an r-form and β is an s-form then the (r+s)-form $\alpha \wedge \beta$ has components

$$(\alpha \wedge \beta)_{i_1 \dots i_{r+s}} = \Sigma\, \delta^{j_1 \dots j_r k_1 \dots k_s}_{i_1 \dots \dots \dots i_{r+s}} \alpha_{j_1 \dots j_r} \beta_{k_1 \dots k_s} .$$

If $X \in E_1$ and α is an r-form, the (r-1)-form $X \lrcorner \alpha$ has components

$$(X \lrcorner \alpha)_{i_1 \dots i_{r-1}} = \Sigma\, X^i \alpha_{i i_1 \dots i_{r-1}} .$$

Let $X_{1'}, \dots, X_{n'}$ be another basis of E, and define $J^{j'}_i$ and $J^k_{j'}$ by

$$X_i = \Sigma\, J^{j'}_i X_{j'} ,$$
$$X_{j'} = \Sigma\, J^k_{j'} X_k .$$

Then

$$\Sigma\, J^{j'}_i J^k_{j'} = \delta^k_i , \qquad \Sigma\, J^j_{i'} J^{k'}_j = \delta^{k'}_{i'} .$$

If $u \in E^s_r$ the components of u with respect to the new basis are

$$u^{i_1',\dots,i_r'}_{j_1',\dots,j_s'} = \Sigma\, J^{\ell_1}_{j_1'} \dots J^{\ell_s}_{j_s'} u^{k_1\dots k_r}_{\ell_1\dots\ell_s} J^{i_1'}_{k_1} \dots J^{i_r'}_{k_r} .$$

Proof. The proof is trivial. QED.

Notice that the primed indices do not take values in the set $\{1,\dots,n\}$ but in a disjoint set $\{1',\dots,n'\}$ of the same cardinality. This notation is very convenient, as it makes it impossible to make a mistake in writing the transformation laws.

12. Classical tensor notation

Despite the profusion of indices, the classical tensor notation is frequently quite useful, especially in computations involving contractions. The vector fields over a coordinate neighborhood in a finite dimensional manifold are a free module of finite type, but the module of all vector fields does not in general have a basis. (If it does, the manifold is called parallelizable.) However, it is possible to use the classical tensor notation globally, without any choice of local coordinates, if we make the following conventions.

Let E be an F module, and let $u \in E^s_r$. Consider an expression of the form

(11) $$u^{i_1\dots i_r}_{j_1\dots j_s} .$$

Instead of "i_1",...,"j_s" we may use any other r+s indices, provided they are distinct indices. The upper indices are called contravariant indices, the lower indices are covariant indices. Next we suppose that the contravariant indices are covariant vector fields and the covariant indices are contravariant vector fields. Then we define (11) to be the scalar

$$u^{i_1 \dots i_r}_{j_1 \dots j_s} = u(i_1, \dots, i_r, j_1, \dots, j_s) \; .$$

It would perhaps be better to write $j^1 \dots j^s$, but we don't.) Notice that lthough the indices are required to be distinct indices, the mathematical bjects they denote need not be distinct. (Thus we may have $i_1 = i_2$ as ovariant vector fields although obviously "i_1" $\neq$ "i_2" .) However, for an -form α we make the special <u>convention</u> that

12) $$\alpha_{i_1 \dots i_r} = r!\alpha(i_1, \dots, i_r) , \qquad \alpha \in A^r \; .$$

Now suppose that E is totally reflexive, so that contractions of ensor fields are meaningful. If $u \in E^s_r$ we define

13) $$u^{i_1 \dots i_{\mu-1} a i_{\mu+1} \dots i_r}_{j_1 \dots j_{\nu-1} a j_{\nu+1} \dots j_s} = (C^\mu_\nu u)(i_1, \dots, \hat{i}_\mu, \dots, i_r, j_1, \dots, \hat{j}_\nu, \dots, j_s) \; .$$

nstead of "a" we may use any other index, provided it is distinct from he other indices occurring. An index which occurs precisely twice, once s an upper index and once as a lower index, is called a <u>dummy index.</u> otice that there is no summation sign in (13). This is because nothing is eing summed. (When dealing with components with respect to a basis of a ree module of finite type, we will continue to write summation signs when ummations occur.) We may have more than one dummy index, provided they are ll distinct from each other and the remaining indices, to indicate repeated ontractions. The notation is unambiguous because, from the definition of ontraction, the order in which the contractions are performed is immaterial.

Here are some examples of the use of this notation. In all but the irst example we assume that E is totally reflexive. If $u \in E^s_r$ and $\in E^{s'}_{r'}$ then

$$(u \otimes v)^{i_1 \dots i_r k_1 \dots k_{r'}}_{j_1 \dots j_s \ell_1 \dots \ell_{s'}} = u^{i_1 \dots i_r}_{j_1 \dots j_s} v^{k_1 \dots k_{r'}}_{\ell_1 \dots \ell_{s'}} \; .$$

If $1 \leq \mu \leq r$, $1 \leq \nu \leq s$ and $u \in E^s_r$ then

$$(C^\mu_\nu u)^{i_1 \cdots \hat{i}_\mu \cdots i_r}_{j_1 \cdots \hat{j}_\nu \cdots j_s} = u^{i_1 \cdots i_{\mu-1} a i_{\mu+1} \cdots i_r}_{j_1 \cdots j_{\nu-1} a j_{\nu+1} \cdots j_s} .$$

If $u \in E^{r+s'}_{s+r'}$, $v \in E^s_r$ and $\iota = \iota(r,s,r',s')$ then

$$((\iota u)v)^{k_1 \cdots k_{r'}}_{\ell_1 \cdots \ell_{s'}} = u^{i_1 \cdots i_s k_1 \cdots k_{r'}}_{j_1 \cdots j_r \ell_1 \cdots \ell_{s'}} v^{j_1 \cdots j_r}_{i_1 \cdots i_s} . \tag{14}$$

The notation here is abusive. The right hand side of (14) is not the product of two scalars but is written instead of

$$(u \otimes v)^{i_1 \cdots i_s k_1 \cdots k_{r'}, j_1 \cdots j_r}_{j_1 \cdots j_r \ell_1 \cdots \ell_{s'}, i_1 \cdots i_s} .$$

We will indulge freely in this abuse of notation. Now let Sym_r be the restriction of Sym to E^r. Since Sym is F-linear, $r!\, \mathrm{Sym}_r = \iota(r,0,r,0)\varepsilon$ for a unique tensor ε in E^r_r, and if $u \in E_r$ then

$$(\mathrm{Sym}\ u)^{i_1 \cdots i_r} = \frac{1}{r!} \varepsilon^{i_1 \cdots i_r}_{j_1 \cdots j_r} u^{j_1 \cdots j_r} .$$

The tensor ε may be computed explicitly, and one finds

$$\varepsilon^{i_1 \cdots i_r}_{j_1 \cdots j_r} = \mathrm{perm} < j_\mu, i_\nu > ,$$

where perm denotes the permanent. (The permanent of a square array of scalars is defined in the same way as the determinant except that there are no minus signs.) Similarly, if $\alpha \in E^r$ then

$$(\mathrm{Alt}\ \alpha)_{i_1 \cdots i_r} = \frac{1}{r!} \delta^{j_1 \cdots j_r}_{i_1 \cdots i_r} \alpha_{j_1 \cdots j_r}$$

for a unique tensor δ in E^r_r, and

$$\delta^{j_1 \cdots j_r}_{i_1 \cdots i_r} = \det < i_\mu, j_\nu >$$

where det denotes the determinant. If $\alpha \in A^r$ and $\beta \in A^s$ then (recall (12))

$$(\alpha\wedge\beta)_{i_1\cdots i_{r+s}} = \delta^{j_1\cdots j_r k_1\cdots k_s}_{i_1\cdots\cdots\cdots i_{r+s}}\alpha_{j_1\cdots j_r}\beta_{k_1\cdots k_s}$$

and if $X \in E$ then

$$(X \lrcorner \alpha)_{i_1\cdots i_{r-1}} = X^a\alpha_{ai_1\cdots i_{r-1}} .$$

13. Tensor fields on manifolds

Let p be a point in the C^∞ manifold M. A tangent vector at p is an equivalence class of differentiable mappings $x\colon \mathbb{R} \longrightarrow M$ with $x(0) = p$, where x and y are equivalent in case the coordinates of $x(t)$ and $y(t)$ differ by $o(t)$. One verifies that this condition is independent of the choice of local coordinates, and that addition and multiplication by constants are well-defined on tangent vectors. Thus the set of all tangent vectors at p forms a real vector space M_p, called the tangent space at p. A cotangent vector at p is the dual notion: an equivalence class of differentiable mappings $f\colon M \longrightarrow \mathbb{R}$ with $f(p) = 0$, where f and g are equivalent if $g(q)$ and $g(q)$ differ by little o of the difference in coordinates of q and p. Again, the condition is independent of the choice of local coordinates, and the cotangent vectors form a vector space M'_p which is in a natural way the dual vector space to M_p.

The set $T(M)$ of all tangent vectors at all points of M has a natural structure of C^∞ manifold as does the set $T^*(M)$ of all cotangent vectors. They are called the tangent bundle and cotangent bundle. They are equipped with natural projections onto M, the projections which assign to each vector the point p at which it lives. A C^∞ section of the tangent bundle is called a contravariant vector field or vector field and a C^∞

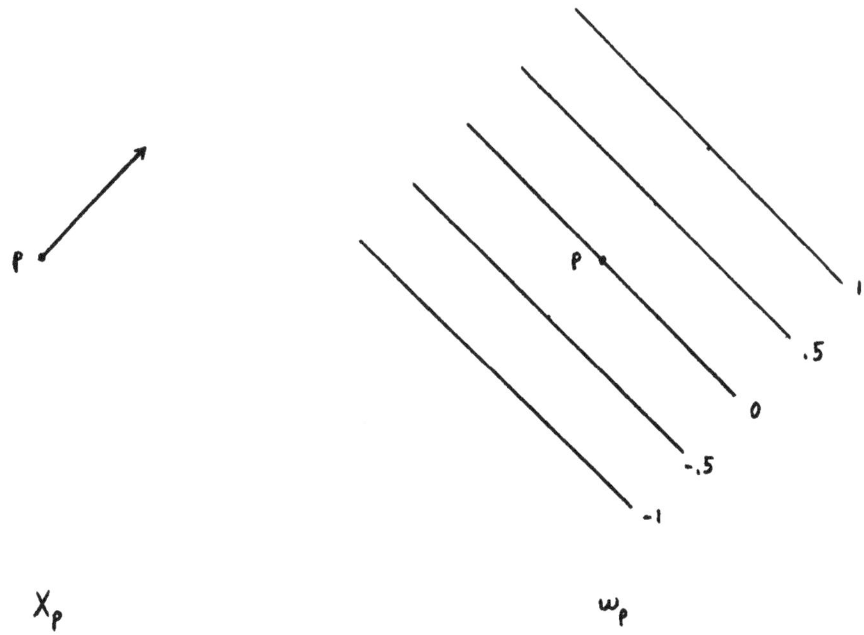

Figure 1. Pictures of a tangent vector X_p and a cotangent vector ω_p. A tangent vector gives a direction and speed of motion, a cotangent vector is a linear approximation to a scalar. The tangent vector $2X_p$ would be indicated by an arrow twice as long, $2\omega_p$ would be indicated a relabeling of the hyperplanes (twice as dense). In the figure X_p and ω_p look as if they are in some sense the same, but this has no meaning unless the tangent space is equipped with additional structure, such as a pseudo-Riemannian metric or symplectic structure.

section of the cotangent bundle is called a covariant vector field or 1-form. They form modules E and E' over the algebra F of all scalars (C^∞ real functions on M). Therefore we have the notions of tensor fields on M and tensors at a point p.

Tensors are of great importance in differential geometry because they are invariantly defined geometrical objects (independent of any coordinate system) which live at points. Both characteristics are necessary in order for an object to be a tensor. Suppose for example we attempt to define a tensor u, contravariant of rank 2, by requiring, in local coordinates,

$$u(\omega,\eta) = \Sigma\, \delta^{ij}\omega_i\eta_j \;,$$

where δ^{ij} is 1 if $i = j$ and 0 otherwise. This lives at points but is not invariantly defined, since in new coordinates $x^{1'},\ldots,x^{n'}$ it would have components

$$\Sigma\, \delta^{k\ell} \frac{\partial x^{i'}}{\partial x^k} \frac{\partial x^{j'}}{\partial x^\ell} \quad .$$

(On the other hand, δ^i_j are in each coordinate system, the components of a certain tensor.) As another example, let X be a fixed contravariant vector field other than 0 and define θ on E by $\theta(Y) = [X,Y]$, where $[X,Y]$ is the Lie product of X and Y (§2). This is invariantly defined but it does not live at points, because in order to know $\theta(Y)$ at a point p we need to know something about Y in a neighborhood of p in order to differentiate it. In fact, θ is $\mathbb{R}$-linear but not F-linear, since $\theta(fY) = f\theta(Y)+(X\cdot f)Y$, so that θ is not a tensor field. The condition of F-linearity is in fact the condition that an $\mathbb{R}$-multilinear object live at points. If for example ω is a 1-form and $X_p = Y_p$ then $(\omega(X))(p) = (\omega(Y))(p)$, since we may write $X-Y = fZ$ with $f(p) = 0$, and so

$$(\omega(X))(p)-(\omega(Y))(p) = (\omega(fZ))(p) = f(p)(\omega(Z))(p) = 0 .$$

The example of the Lie product shows that not all interesting geometrical objects are tensors. Affine connections are another example of second-order geometrical objects. Tensor fields are first-order geometrical objects since the notion of tangent vector involves one derivative.

14. Tensors and mappings

Suppose we have two F^0 algebras F and $\tilde{F}$, an F module E with dual E', and an $\tilde{F}$ module $\tilde{E}$ with dual $\tilde{E}'$. We shall use the word homomorphism for any of the following:

(15) $$\rho\colon F \longrightarrow \tilde{F} ,$$

an F^0 algebra homomorphism;

(16) $$\rho\colon E \longrightarrow \tilde{E} ,$$

a group homomorphism (and similarly for $\rho\colon E' \longrightarrow \tilde{E}'$) ;

(17) $$\rho\colon (F,E) \longrightarrow (\tilde{F},\tilde{E})$$

where $\rho\colon F \longrightarrow \tilde{F}$ and $\rho\colon E \longrightarrow \tilde{E}$ are homomorphisms satisfying the compatability condition

(18) $$\rho(fX) = \rho(f)\rho(X) , \qquad f \in F,\ X \in E ,$$

(and similarly for $\rho\colon (F,E') \longrightarrow (\tilde{F},\tilde{E}')$) ; and finally for

(19) $$\rho\colon (F,E,E') \longrightarrow (\tilde{F},\tilde{E},\tilde{E}')$$

where $\rho\colon F \longrightarrow \tilde{F}$, $\rho\colon E \longrightarrow \tilde{E}$, $\rho\colon E' \longrightarrow \tilde{E}'$ are homomorphisms satisfying the compatability conditions (18) and

(20) $$\rho(f\omega) = \rho(f)\rho(\omega) , \qquad f \in F,\ \omega \in E' ,$$

(21) $$\rho(<\omega,X>) = <\rho(\omega),\rho(X)>, \qquad \omega \in E', X \in E .$$

Now let

$$\Phi: M \longrightarrow \tilde{M}$$

be a C^∞ mapping of the manifold M into the manifold $\tilde{M}$. Then

$$\Phi^*: \tilde{F} \longrightarrow F$$

defined by $(\Phi^*\tilde{f})(p) = \tilde{f}(\Phi(p))$ is a homomorphism. If we recall what a tangent vector at a point p in M is, we see that Φ induces a vector space homomorphism (linear transformation)

$$d\Phi_p: M_p \longrightarrow \tilde{M}_{\Phi(p)} .$$

It is called the differential of Φ at p. By duality,

$$(d\Phi_p)' : \tilde{M}'_{\Phi(p)} \longrightarrow M'_p .$$

If we define

$$<(\Phi^*\tilde{\omega})_p, X_p> = <\tilde{\omega}_{\Phi(p)}, d\Phi_p(X_p)>$$

then

$$\Phi^*: (\tilde{F},\tilde{E}') \longrightarrow (F,E')$$

is a homomorphism. In the same way we obtain a homomorphism

$$\Phi^*: (\tilde{F},\tilde{E}^*) \longrightarrow (F,E^*)$$

of the covariant tensor algebras, which preserves the grading and products $\otimes$, and sends the Grassmann algebra $\tilde{A}^*$ into A^*. However, it is immediately clear to almost anyone that we do not in general obtain a homomorphism $(F,E) \longrightarrow (\tilde{F},\tilde{E})$ since Φ is not necessarily onto, we may not have $d\Phi_p(X_p) = d\Phi_q(X_q)$ whenever $\Phi(p) = \Phi(q)$, and even if these difficulties do not arise we may not get a C^∞ section of $T(\tilde{M})$ (see Exercise A.4 on p.83 of Helgason [2]). The mapping Φ induces C^∞ maps

$$\Phi_*: T(M) \longrightarrow T(\tilde{M}) ,$$
$$\Phi^*: T^*(\tilde{M}) \longrightarrow T^*(M) ,$$

but Φ_* does not in general induce a mapping on C^∞ sections of $T(M)$.

Suppose now that Φ is a diffeomorphism of M onto $\tilde{M}$. Then we obtain a homomorphism (in fact, an isomorphism)

$$\Phi^*: (\tilde{F},\tilde{E},\tilde{E}') \longrightarrow (F,E,E') \tag{22}$$

as follows. On $\tilde{F}$ and $\tilde{E}'$, Φ^* is as defined above. For $\tilde{X}$ in $\tilde{E}$ we define

$$(\Phi^*\tilde{X})(p) = d(\Phi^{-1})_{\Phi(p)}\tilde{X}_{\Phi(p)} .$$

This homomorphism extends in a natural way to the mixed tensor algebras. In the same way we obtain a homomorphism (22) if Φ is an imbedding of M in $\tilde{M}$.

It is unfortunate that covariant tensor fields transform contravariantly under point mappings of manifolds, but it is too late to change the terminology. Early geometers were more concerned with coordinate changes than point mappings, and coordinates are scalars, which transform the same way as covariant tensor fields.

Notice that we have used the notation E^* for covariant tensor fields in keeping with the fact that they transform the opposite way to point mappings. For example, the cohomology ring is formed from the Grassmann algebra A^* and it is universally denoted H^*.

In our study of tensor analysis we shall make no use of points except at one point in the discussion of harmonic forms (§7), where we will need the following notion.

__Definition.__ The F module E is __punctual__ if there exists a separating family of homomorphisms of the form

$$\rho_p: (F,E,E') \longrightarrow (F_p,E_p,E'_p)$$

where $F_p = F^0$ and E_p is a finite dimensional F^0 vector space.

The module of contravariant vector fields on a manifold is punctual: take ρ_p to be evaluation at the point p and E_p to be the tangent space at p.

References

[1] N. Bourbaki, Eléments de mathématique, Hermann, Paris. See especially Book 2, Algèbre, Chaps. 2 and 3.

[2] Sigurður Helgason, Differential Geometry and Symmetric Spaces, Academic Press, New York, 1962.

§2. Derivations on scalars

1. Lie products

A derivation of F is an F^0-linear mapping $X: F \longrightarrow F$ such that

$$X(fg) = (Xf)g + f(Xg) , \qquad f,g \in F .$$

(See §1.1 for the assumptions on F^0 and F .)

If X and Y are derivations then so is $X+Y$ defined by

$$(X+Y)f = Xf + Yf ,$$

and if h is in F then hX defined by

$$(hX)f = h(Xf)$$

is also a derivation. Thus the set of all derivations of F is an F module. It is denoted D .

If $X \in D$ then $X1 = X1 + X1$, so that $X1 = 0$. By F^0-linearity, $Xa = 0$ for all a in F^0 .

If X and Y are in D , we define their Lie product $[X,Y]$ by

$$[X,Y]f = XYf - YXf .$$

This is again a derivation:

$$\begin{aligned}[X,Y](fg) &= XY(fg) - YX(fg) \\ &= X\{(Yf)g + f(Yg)\} - Y\{(Xf)g + f(Xg)\} \\ &= (XYf)g + (Yf)(Xg) + (Xf)(Yg) + f(XYg) \\ &\quad - (YXf)g - (Xf)(Yg) - (Yf)(Xg) - f(YXg) \\ &= ([X,Y]f)g + f([X,Y]g) .\end{aligned}$$

The set of all F^0-linear mappings of F into itself forms an associative ring, and D is a subset of it. In any associative ring we define the Lie product of any two elements X and Y to be $[X,Y] = XY - YX$.

A simple computation shows that the Jacobi identity

$$[[X,Y],Z] + [[Y,Z],X] + [[Z,X],Y] = 0 \tag{1}$$

holds in any associative ring. Since D is closed under the formation of Lie products, the Jacobi identity holds in it. (However, with respect to the Lie product as multiplication D is not in general associative.) The Jacobi identity may be rewritten as

$$[X,[Y,Z]] - [Y,[X,Z]] = [[X,Y],Z] . \tag{2}$$

Define θ_X on D by

$$\theta_X Y = [X,Y] . \tag{3}$$

Then the Jacobi identity (2) is

$$[\theta_X, \theta_Y] = \theta_{[X,Y]} . \tag{4}$$

More generally, if Y is an iterated Lie product of n elements $X_1, \ldots, X_n$, associated in any way, then by induction $\theta_W Y$ is the sum of n terms, in the μ-th of which X_μ is replaced by $\theta_W X_\mu$. For example,

$$[W,[[X,Y],Z]] = [[[W,X],Y],Z] + [[X,[W,Y]],Z] + [[X,Y],[W,Z]] . \tag{5}$$

If X and Y are in D then clearly

$$[Y,X] = -[X,Y] . \tag{6}$$

The Lie product is F^0-bilinear, so that with the Lie product as multiplication D is an F^0 algebra (not in general associative). An F^0 algebra satisfying (6) and (1) is called a Lie algebra, so that D is a Lie algebra over F^0.

However, the Lie product is not F-bilinear. In fact,

$$[fX, gY] = fg[X,Y] + f(Xg)Y - g(Yf)X . \tag{7}$$

2. Lie modules

Definition. A Lie module is an F module E together with an F-linear mapping $X \longrightarrow X\cdot$ of E into derivations of F and a mapping $(X,Y) \longrightarrow [X,Y]$ of $E \times E$ into E such that with respect to it E is a Lie algebra over F^0 and

(8) $$[X,fY] = f[X,Y] + (X\cdot f)Y\,,$$

(9) $$[X,Y]\cdot f = X\cdot Y\cdot f - Y\cdot X\cdot f$$

for all f in F and X,Y in E.

Notice that from (8) and the fact that $[Y,X] = -[X,Y]$ (since E is a Lie algebra over F^0) the more general relation

(10) $$[fX,gY] = fg[X,Y] + f(X\cdot g)Y - g(Y\cdot f)X$$

holds. We have the following.

Theorem 1. Let D be the module of all derivations of F, with $X\cdot f = Xf$ and $[X,Y] = XY-YX$. Then D is a Lie module.

The main example of interest is the Lie module E of all derivations of the algebra F of C^∞ functions on a manifold, which may be identified with the set of all contravariant vector fields. In the definition of Lie module we did not assume that the mapping $X \longrightarrow X\cdot$ is injective. Other examples of a Lie module are a Lie algebra over F^0 when $F = F^0$, and the set of all vector fields on a manifold invariant under some C^∞ action of a Lie group over the algebra F of all invariant scalars.

Roughly speaking, a Lie module is like a Lie algebra except that the elements of the module act on the coefficients by derivations in a natural way.

Definition. Let E be a Lie module, f a scalar. The differential of f, df, is defined by

$$(df)(X) = X\cdot f\ , \qquad X \in E\ .$$

Since $X \longrightarrow X\cdot$ is F-linear, the differential of f is a 1-form (§1.2).

3. Coordinate Lie modules

Definition. A Lie module E is called a coordinate Lie module in case there exist scalars $x^1,\ldots,x^n$ (called coordinates) whose differentials are a basis for the module of 1-forms.

Let E be a coordinate Lie module with coordinates $x^1,\ldots,x^n$. Then E^1, and consequently $E = E_1$, is free of finite type (§1.11) and therefore totally reflexive (§1.5). The

$$dx^1,\ldots,dx^n \tag{11}$$

are a basis of E^1. The dual basis, which is a basis of E, is denoted

$$\frac{\partial}{\partial x^1},\ldots,\frac{\partial}{\partial x^n} \tag{12}$$

so that

$$< dx^i, \frac{\partial}{\partial x^j} > = \delta^i_j\ .$$

Observe that the elements of the basis (12) commute (i.e. their Lie products are 0) since

$$< dx^k, [\frac{\partial}{\partial x^i}, \frac{\partial}{\partial x^j}]> = \frac{\partial}{\partial x^i}\frac{\partial}{\partial x^j}x^k - \frac{\partial}{\partial x^j}\frac{\partial}{\partial x^i}x^k$$

$$= \frac{\partial}{\partial x^i} < dx^k, \frac{\partial}{\partial x^j} > - \frac{\partial}{\partial x^j} < dx^k, \frac{\partial}{\partial x^i} > = \frac{\partial}{\partial x^i}\delta^k_j - \frac{\partial}{\partial x^j}\delta^k_i = 0\ ,$$

due to the fact that the δ^k_j, δ^k_i are constants (1 or 0). Notice that the symbol $\partial/\partial x^1$, for example, has meaning only if the entire coordinate system $x^1,\ldots,x^n$ is given (in contrast to dx^1, which is simply the differential of the scalar x^1). Thus if x^1,x^2 are coordinates so are $x^{1'}=x^1$, $x^{2'} = x^2+x^1$ and $\partial/\partial x^{1'} \neq \partial/\partial x^1$ even though $x^{1'} = x^1$.

Let f be a scalar, let X be a vector field with components X^i, and let Y be a vector field with components Y^i. Then df has components

$$\frac{\partial f}{\partial x^i},$$

$X\cdot f$ is the scalar

$$\Sigma\, X^i \frac{\partial f}{\partial x^i},$$

and $[X,Y]$ has components

$$\Sigma(X^a \frac{\partial}{\partial x^a} Y^i - Y^a \frac{\partial}{\partial x^a} X^i).$$

If $x^{1'},\ldots,x^{n'}$ are also coordinates then the Jacobian matrix $J^{j'}_i$ of §1.11 is

$$J^{j'}_i = \frac{\partial x^{j'}}{\partial x^i}$$

and similarly for its inverse, so that we have the familiar formulas

$$dx^{i'} = \Sigma \frac{\partial x^{i'}}{\partial x^j}\, dx^j,$$

$$\frac{\partial}{\partial x^{i'}} = \Sigma \frac{\partial x^j}{\partial x^{i'}} \frac{\partial}{\partial x^j}.$$

4. Vector fields and flows

Let us discuss heuristically but in some detail why contravariant vector fields on a manifold are the same as derivations of the algebra of scalars, and what the geometrical meaning of scalar multiples, sums, and Lie products is.

Let X be a (contravariant) vector field on a manifold M, everything being assumed to be of class C^∞. Thus X is an assignment of velocities at each point p of the manifold. The fundamental existence theorem for differential equations shows how to integrate to obtain the motion of a particle starting at any point. In this way we obtain (at least if M is compact) a one-parameter family of diffeomorphisms $\Phi(t)$ of M onto itself such that $\Phi(t+s) = \Phi(t)\Phi(s)$. We call this a flow, and in this discussion we shall ignore the fact that in general $\Phi(t)$ is only defined locally if M is not compact. Thus $\Phi(t)p$ is the position of a particle at time t if it starts at $\Phi(0)p = p$ at time 0. As discussed in §1.14, if f is a scalar (C^∞ function) we have $\Phi(t)^*f$ given by $(\Phi(t)^*f)(p) = f(\Phi(t)p$. But $\Phi(t)p$ is a representative of the tangent vector X_p at p (see §1.13), so that the derivative of $(\Phi(t)^*f)(p)$ at $t = 0$ is known if X_p is known. We let

$$(Xf)(p) = \lim_{t\to 0} \frac{(\Phi(t)^*f)(p) - f(p)}{t} .$$

This gives the action of vector fields on scalars. It is perhaps worth recalling why this is a derivation:

$$(X(fg))(p) = \lim_{t\to 0} \frac{(fg)(\Phi(t)p) - (fg)(p)}{t}$$

$$= \lim_{t\to 0} \{\frac{f(\Phi(t)p)-f(p)}{t} \frac{g(\Phi(t)p)+g(p)}{2} + \frac{f(\Phi(t)p) + f(p)}{2} \frac{g(\Phi(t)p)-g(p)}{t}\}$$

$$= \{(Xf)g + f(Xg)\}(p) .$$

Conversely, if X is a derivation on the scalars and p is a point then the $f(p) + t(Xf)(p)$, where f is allowed to be $x^1, \ldots, x^n$ for a set of local coordinates at p, determines a curve whose equivalence class (the tangent vector at p) is independent of the choice of local coordinates. Consequently we identify the vector fields on a manifold with the derivations of the algebra of scalars. Notice that the assumption that everything is of class C^∞ is necessary for this identification.

Now we shall discuss the meaning of scalar multiplication, addition, and Lie products in terms of flows.

Let X generate the flow Φ, let h be a scalar, and let hX generate the flow Ψ. Then Ψ is the same as Φ except for a change of time scale. The new velocity is hX, so we have (letting s be the new time parameter)

$$\frac{dt}{ds} = h(\Phi(t)p) \ .$$

Thus

$$\Psi(s)p = \Phi(t)p$$

where

$$s = \int_0^t \frac{1}{h(\Phi(r)p)}\, dr \ .$$

If $h(p) = 0$ then $\Psi(s)p = p$.

Now let X generate Φ, Y generate Ψ, and $X+Y$ generate Ω. If X and Y commute (i.e., if $[X,Y] = 0$) then Φ and Ψ commute (locally) and Ω is given by

$$\Omega(t)p = \Phi(t)\Psi(t)p \ .$$

The general case is more complicated, and

$$\Omega(t)p = \lim_{n\to\infty} \left(\Phi\left(\frac{t}{n}\right)\Psi\left(\frac{t}{n}\right)\right)^n p \ . \tag{13}$$

Roughly speaking, the flow Ω is the simultaneous action of the flows Φ and Ψ. To see (13) formally, notice that formally

$$\Phi(t)^* = e^{tX},$$

since $\Phi(t+s)^* = \Phi(t)^*\Phi(s)^*$ and

$$\frac{d}{dt}\,\Phi(t)^*\Big|_{t=0} = X .$$

Then the expansion of

$$\left(e^{\frac{t}{n}X}\,e^{\frac{t}{n}Y}\right)^n$$

in powers of t is the same as the expansion of $e^{t(X+Y)}$ in powers of t, except for a fraction of terms in each order such that the fraction tends to 0 as $n \longrightarrow \infty$. The product formula (13) in the very general setting of semigroups on Banach spaces is due to Trotter [3].

Finally, let X generate the flow Φ, Y generate the flow Ψ, and $[X,Y]$ generate the flow Θ. Then

$$\Theta(t)p = \lim_{n\to\infty}\left(\Psi(-\sqrt{\tfrac{t}{n}})\Phi(-\sqrt{\tfrac{t}{n}})\Psi(\sqrt{\tfrac{t}{n}})\Phi(\sqrt{\tfrac{t}{n}})\right)^n p . \tag{14}$$

To see (14) formally, we make a formal computation to second order in t;

$$\begin{aligned} &e^{tX}e^{tY}e^{-tX}e^{-tY} \\ &\quad = (1+tX+\tfrac{t^2}{2}X^2)(1+tY+\tfrac{t^2}{2}Y^2)(1-tX+\tfrac{t^2}{2}X^2)(1-tY+\tfrac{t^2}{2}Y^2)+o(t^2) \\ &\quad = 1+t^2X^2+t^2Y^2+t^2XY-t^2X^2-t^2XY-t^2YX-t^2Y^2+t^2XY+o(t^2) \\ &\quad = 1+t^2[X,Y]+o(t^2) . \end{aligned}$$

If we replace t^2 by t this gives, formally,

$$\begin{aligned} e^{t[X,Y]} &= e^{\sqrt{t}X}e^{\sqrt{t}Y}e^{-\sqrt{t}X}e^{-\sqrt{t}Y}+o(t) \\ &= \lim_{n\to\infty}\left(e^{\sqrt{\frac{t}{n}}X}\,e^{\sqrt{\frac{t}{n}}Y}\,e^{-\sqrt{\frac{t}{n}}X}\,e^{-\sqrt{\frac{t}{n}}Y}\right)^n . \end{aligned}$$

This computation concerns the action of the flows on scalars, and the result is equivalent to (14) for the point flows. Helgason essentially gives a proof of (14) for the case when X and Y are in the Lie algebra of a Lie group (see pages 96, 97 and 105 of [2, §1]). It should not be difficult to establish (14) in the general case of vector fields on a manifold.

We conclude with an example. Consider a car. The configuration space of a car is the four dimensional manifold

$$M = \mathbb{R}^2 \times \mathbb{T}^2$$

parameterized by (x,y,φ,θ), where (x,y) are the Cartesian coordinates of the center of the front axle, the angle φ measures the direction in which the car is headed, and θ is the angle made by the front wheels with the car. (More realistically, the configuration space is the open submanifold $-\theta_{max} < \theta < \theta_{max}$.) See Figure 2.

There are two distinguished vector fields, called Steer and Drive, on M corresponding to the two ways in which we can change the configuration of a car. Clearly

(15) $$\text{Steer} = \frac{\partial}{\partial\theta}$$

since in the corresponding flow θ changes at a uniform rate while x,y and φ remain the same. To compute Drive, suppose that the car, starting in the configuration (x,y,φ,θ), moves an infinitesimal distance h in the direction in which the front wheels are pointing. In the notation of Figure 3,

$$D = (x+h\cos(\varphi+\theta)+o(h),\ y+h\sin(\varphi+\theta)+o(h)) .$$

Let $\ell = \overline{AB}$ be the length of the tie rod (if that is the name of the

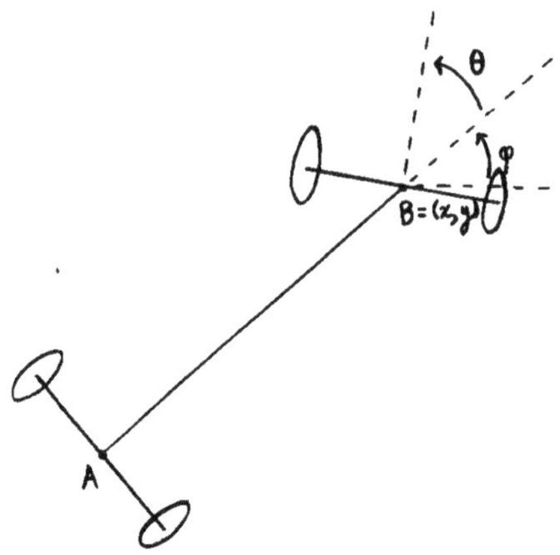

Figure 2. A car

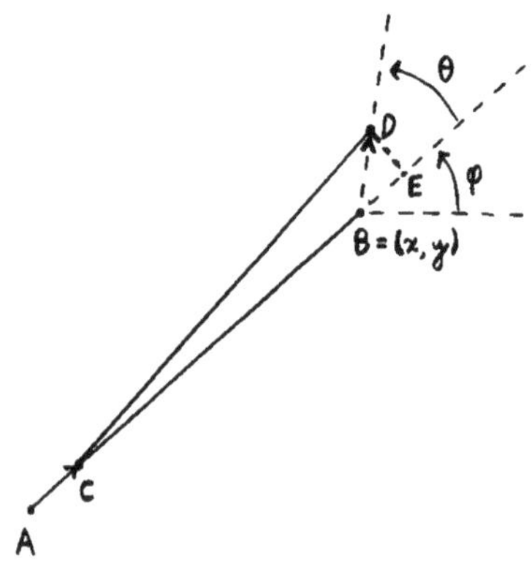

Figure 3. A car in motion

thing connecting the front and rear axles). Then $\overline{CD} = \ell$ too since the tie rod does not change length (in non-relativistic mechanics). It is readily seen that $\overline{CE} = \ell + o(h)$, and since $\overline{DE} = h \sin \theta + o(h)$ the angle BCD (which is the increment in φ) is $h \sin \theta / \ell$, while θ remains the same. Let us choose units so that $\ell = 1$. Then

$$\text{Drive} = \cos(\varphi+\theta) \frac{\partial}{\partial x} + \sin(\varphi+\theta) \frac{\partial}{\partial y} + \sin \theta \frac{\partial}{\partial \varphi}. \tag{16}$$

By (15) and (16),

$$[\text{Steer}, \text{Drive}] = -\sin(\varphi+\theta) \frac{\partial}{\partial x} + \cos(\varphi+\theta) \frac{\partial}{\partial y} + \cos \theta \frac{\partial}{\partial \varphi} . \tag{17}$$

Let

$$\text{Slide} = -\sin \varphi \frac{\partial}{\partial x} + \cos \varphi \frac{\partial}{\partial y} ,$$
$$\text{Rotate} = \frac{\partial}{\partial \varphi} .$$

Then the Lie product of Steer and Drive is equal to Slide + Rotate on $\theta = 0$, and generates a flow which is the simultaneous action of sliding and rotating. This motion is just what is needed to get out of a tight parking spot. By formula (14) this motion may be approximated arbitrarily closely, even with a restriction $-\theta_{max} < \theta < \theta_{max}$ with θ_{max} arbitrarily small, in the following way: steer, drive, reverse steer, reverse drive, steer, drive, reverse steer,... . What makes the process so laborious is the square roots in (14).

Let us denote the Lie product (17) of Steer and Drive by Wriggle. Then further simple computations show that we have the commutation relations

$$[\text{Steer}, \text{Drive}] = \text{Wriggle},$$
$$[\text{Steer}, \text{Wriggle}] = -\text{Drive},$$
$$[\text{Wriggle}, \text{Drive}] = \text{Slide},$$

and the commutator of Slide with Steer, Drive, and Wriggle is zero. Thus the four vector fields span a four dimensional solvable Lie algebra over $\mathbb{R}$.

To get out of an extremely tight parking spot, Wriggle is insufficient because it may produce too much rotation. The last commutation relation shows, however, that one may get out of an arbitrarily tight parking spot in the following way: wriggle, drive, reverse wriggle (this requires a cool head), reverse drive, wriggle, drive,... .

The example illustrates a phenomenon of frequent occurrence in differential geometry, namely holonomy, or rather the lack of holonomy. The vector fields Steer and Drive, which at first sight give the only possible motions of a car, span a module over the scalars which is not closed under the formation of Lie products. That is, the field of two dimensional planes in the tangent bundle is not integrable (not involutive, not holonomic) and so is not the field of tangent planes to a family of two dimensional surfaces. Motions which at first sight are impossible can in fact be approximated arbitrarily closely (in the C^0 topology but not the C^1 topology) by possible motions.

Reference

[3] H. F. Trotter, On the product of semi-groups of operators, Proceedings of the American Mathematical Society 10(1959), 545-551.

§3. Derivations on tensors

1. Algebra derivations

Let K be an F^0 algebra, not necessarily commutative or associative. A <u>derivation</u> X of K is an F^0-linear mapping of K into itself such that $X(uv) = (Xu)v + u(Xv)$ for all u and v in K. The computation in §2.1 used neither commutativity nor associativity, so if X and Y are derivations of K so is $[X,Y]$. The derivations lie in the associative algebra of endomorphisms of K as an F^0 vector space, so the Jacobi identity holds. Thus the derivations of K form a Lie algebra over F^0.

Now suppose that K is a graded algebra. That is, K is the weak direct sum

$$K = \sum_{r=-\infty}^{\infty} K_r$$

where each $K_r K_s \subset K_{r+s}$. The K_r with $r < 0$ are usually but not necessarily 0. An F^0 linear mapping X of K into itself is <u>homogeneous of degree</u> a if each $XK_r \subset K_{r+a}$, and <u>homogeneous</u> if it is homogeneous of degree a for some a. The notions of a bi-graded algebra, and bi-homogeneous mappings of bi-degree (a,b), are defined similarly. An <u>antiderivation</u> of a graded algebra K is an F^0-linear mapping of K into itself such that

$$X(uv) = (Xu)v + (-1)^r u(Xv)\,, \qquad u \in K_r,\ v \in K\,.$$

The <u>anticommutator</u> of X and Y is $XY + YX$. A simple calculation establishes the following theorem.

<u>Theorem 1.</u> <u>Let</u> X <u>and</u> Y <u>be antiderivations on the graded algebra</u> K, <u>homogeneous of odd degrees</u> a <u>and</u> b <u>respectively. Then the anticommutator</u> $XY + YX$ <u>is a derivation of</u> K, <u>homogeneous of degree</u> $a+b$.

2. Module derivations

Let E be an F module. In §1.14 we defined the notion of a homomorphism of (F,E) and consequently we have the notion of an automorphism of (F,E). Formally, let $\rho(t)$ be a one-parameter group of automorphisms of (F,E) and let φ be the derivative of ρ at $t=0$. (For example, $\rho(t)$ may be $\Phi(t)^*$ where $\Phi(t)$ is a flow on a manifold.) By the product rule for differentiation we obtain, formally,

$$\varphi(fX) = f\varphi(X) + \varphi(f)X \,, \qquad f \in F,\quad X \in E \,. \tag{1}$$

This motivates the following definition.

Definition. Let E be an F module. A derivation φ of (F,E) is a derivation φ of F and an F°-linear mapping $\varphi\colon E \longrightarrow E$ such that (1) holds. A *derivation* φ of (F,E,E') is a derivation φ of (F,E) and an F°-linear mapping $\varphi\colon E' \longrightarrow E'$ such that in addition

$$\varphi(f\omega) = f\varphi(\omega) + \varphi(f)\omega \,, \qquad f \in F,\quad \omega \in E' \,, \tag{2}$$

$$\varphi\langle \omega, X\rangle = \langle \varphi,\omega X\rangle + \langle \omega, \varphi X\rangle \,, \qquad \omega \in E',\quad X \in E \,. \tag{3}$$

The motivation for (2) and (3) is again the product rule for differentiation, as it is of course the motivation for the definition of derivation of an F° algebra. We shall find it convenient to indicate derivations of a number of different F modules by the same symbol φ. This is legitimate, provided of course they all give the same derivation of F, since we may regard φ as defined on the disjoint union of F and the various F modules.

Theorem 2. Let E *be an* F *module and let* φ *be a derivation of* (F,E). *Then:*

(a) φ *has a unique extension to a derivation* φ *of* (F,E,E').

(b) *If we extend* φ *to* E'' *then* $\varphi(\kappa X) = \kappa\varphi(X)$ *for all* X *in* E *, where* κ *is the natural mapping of* E *into* E'' .

(c) φ *on* (F,E,E') *has a unique extension which is a derivation* φ *of* $\overset{\circ *}{E}_*$ *as an* F° *algebra, and* φ *is a derivation on* $(F,\overset{\circ}{E}{}^s_r)$ *for all* r *and* s .

(d) *If we extend* φ *to be a derivation on* $(F,\overset{\circ}{E}{}^s_r,E^r_s)$ *and define* φ *on all of* E^*_* *by additivity, then* φ *is a derivation of* E^*_* *as an* F° *algebra. For all* u *in* E^s_r ,

$$(4) \qquad \varphi(u(\omega^1,\ldots,\omega^r,X_1,\ldots,X_s)) = (\varphi u)(\omega^1,\ldots,\omega^r,X_1,\ldots,X_s)$$
$$+ \sum_{\mu=1}^{r} u(\omega^1,\ldots,\varphi\omega^{\mu},\ldots,\omega^r,X_1,\ldots,X_s)$$
$$+ \sum_{\mu=1}^{s} u(\omega^1,\ldots,\omega^r,X_1,\ldots,\varphi X_{\mu},\ldots,X_s) \ .$$

Proof. We define φ on E' by (3). Then each $\varphi\omega$ is a 1-form since if we replace X by fX the additional terms $\varphi(f)\langle\omega,X\rangle$ on the two sides of (3) cancel. Similarly, (2) holds, so that φ is a derivation of (F,E,E') . The uniqueness is clear.

By (a) we do have a unique extension of φ to E'' (such that it is a derivation of $(F,E',E''))$. For all X in E ,

$$\varphi\langle\omega,\kappa X\rangle = \langle\varphi\omega,\kappa X\rangle + \langle\omega,\varphi(\kappa X)\rangle \ .$$

By definition of κ , κ may be dropped from the first two terms of this, and by comparison with (3) we see that $\langle\omega,\varphi(\kappa X)\rangle = \langle\omega,\varphi X\rangle = \langle\omega,\kappa(\varphi X)\rangle$ for all ω in E' , so that $\varphi(\kappa X) = \kappa(\varphi X)$.

The uniqueness assertion in (c) is clear since F, E, and E' generate $\overset{\circ *}{E}_*$ as an F° algebra. To prove existence, we need only show that if H and K are F modules, φ a derivation on (F,H) and φ a derivation on (F,K) (agreeing on F), then

$$(5) \qquad \varphi(X\otimes Y) = \varphi X\otimes Y + X\otimes \varphi Y$$

is well-defined on $H\otimes K$ and extends by additivity to a derivation of $(F,H\otimes K)$. To see this, notice that (5) is obviously well-defined on the free Abelian group used in the definition of tensor product (§1.3) and that φ sends $(fX)\otimes Y - X\otimes(fY)$ into the subgroup generated by the relations imposed in §1.3, and so is well-defined on $H\otimes K$. It is then clear that φ is a derivation on $(F,H\otimes K)$.

By (a) and (c) and the fact (§1.6) that E^r_s is the dual of $\overset{o}{E}{}^s_r$, φ has a unique extension as a derivation of $(F,\overset{o}{E}{}^s_r,E^r_s)$. We extend φ by additivity to E^*_*. Let $u \in E^s_r$, $v \in E^{s'}_{r'}$, $y \in \overset{o}{E}{}^r_s$, $z \in \overset{o}{E}{}^{r'}_{s'}$. Then, by (3),

$$\varphi\langle u\otimes v, y\otimes z\rangle = \langle \varphi(u\otimes v), y\otimes z\rangle + \langle u\otimes v, \varphi y\otimes z + y\otimes\varphi z\rangle ,$$

so that

$$((\varphi u)(y))v(z) + u(\varphi y)v(z) + u(y)((\varphi v)(z)) + u(y)v(\varphi z)$$

$$= \langle \varphi(u\otimes v), y\otimes z\rangle + u(\varphi y)v(z) + u(y)v(\varphi z) .$$

That is, $\varphi u\otimes v + u\otimes\varphi v = \varphi(u\otimes v)$, so that φ is a derivation of E^*_* as an F^o algebra. The final formula (4) is simply (3) for u in E^s_r and

$$\omega^1\otimes\ldots\otimes\omega^r\otimes X_1\otimes\ldots\otimes X_s$$

in $\overset{o}{E}{}^r_s$. QED.

Notice that by (b) if E is totally reflexive, so that we may identify $\overset{o}{E}{}^s_r$ and E^s_r, the two definitions of φ agree. The various derivations given by Theorem 2 will be called the derivations <u>induced</u>

by the derivation φ of (F,E) .

Theorem 3. Let E *be an* F *module and let* φ_1, φ_2 *be derivations of* (F,E) . *Then the commutator* $[\varphi_1, \varphi_2]$ *of* φ_1 *and* φ_2 *is a derivation of* (F,E) . *The commutators of the derivations induced by* φ_1 *and* φ_2 *are the derivations induced by* $[\varphi_1, \varphi_2]$.

Proof. We have

$$\begin{aligned}\varphi_1\varphi_2(fX) &= \varphi_1\{f\varphi_2(X) + \varphi_2(f)X\} \\ &= f\varphi_1\varphi_2(X) + \varphi_1(f)\varphi_2(X) + \varphi_2(f)\varphi_1(X) + \varphi_1(\varphi_2(f))X ,\end{aligned}$$

and similarly for $\varphi_2\varphi_1(fX)$, so that

$$[\varphi_1,\varphi_2](fX) = f[\varphi_1,\varphi_2](X) + ([\varphi_1,\varphi_2]f)X ,$$

and we know that $[\varphi_1,\varphi_2]$ is a derivation of F . The last statement of the theorem is an immediate consequence of the uniqueness assertions in Theorem 2. QED.

By Theorem 2 and §1.6, if φ is a derivation of (F,E) , $u \in E^{r+s'}_{s+r'}$, and $\iota = \iota(r,s,r',s')$ we have the following diagram:

$$\begin{array}{ccc} \mathring{E}^s_r & \xrightarrow{\ \iota u\ } & E^{s'}_{r'} \\ \varphi\big\downarrow & & \big\downarrow\varphi \\ \mathring{E}^s_r & \xrightarrow{\ \iota u\ } & E^{s'}_{r'} \end{array}$$

In general it does not commute, and $[\varphi, \iota u] = \varphi\circ(\iota u) - (\iota u)\circ\varphi$ is not 0 .

Theorem 4. Let E *be an* F *module,* φ *a derivation of* (F,E) , $u \in E^{r+s'}_{s+r'}$, *and* $\iota = \iota(r,s,r',s')$. *Then*

(6) $$[\varphi, \iota u] = \iota\varphi u .$$

Proof. By the definition (§1.6) of ι, if $y \in \overset{\circ}{E}{}^{s}_{r}$, $z \in \overset{\circ}{E}{}^{r}_{s}$, and $v = (\iota u)(y)$ then $v(z) = u(y \otimes z)$. Therefore, by (4),

$$\begin{aligned}(\varphi v)(z) &= \varphi(v(z)) - v(\varphi z) \\ &= \varphi(u(y \otimes z)) - u(y \otimes \varphi z) \\ &= (\varphi u)(y \otimes z) + u(\varphi y \otimes z)\end{aligned}$$

so that (6) holds. QED.

3. Lie derivatives

Suppose that E is a Lie module (§2.2). By the definition of Lie module, if $X \in E$ and we let

$$\theta_X f = X \cdot f \,, \qquad f \in F \,,$$

$$\theta_X Y = [X, Y] \,, \qquad Y \in E \,,$$

then θ_X is a derivation of (F, E). The induced derivation θ_X on the mixed tensor algebra is called the Lie derivative. Thus θ_X is defined on 1-forms ω by (3), which gives

(7) $$X \cdot \langle \omega, Y \rangle = \langle \theta_X \omega, Y \rangle + \langle \omega, [X, Y] \rangle \,, \qquad Y \in E$$

and for tensors u contravariant of rank r and covariant of rank s by

(8) $$\begin{aligned} X \cdot u(\omega^1, \ldots, \omega^r, X_1, \ldots, X_s) &= (\theta_X u)(\omega^1, \ldots, \omega^r, X_1, \ldots, X_s) \\ &+ \sum_{\mu=1}^{r} u(\omega^1, \ldots, \theta_X \omega^\mu, \ldots, \omega^r, X_1, \ldots, X_s) \\ &+ \sum_{\mu=1}^{s} u(\omega^1, \ldots, \omega^r, X_1, \ldots, [X, X_\mu], \ldots, X_s) \,. \end{aligned}$$

If X is a vector field on a manifold, generating the flow $\Phi(t)$, then for any tensor field u, $\theta_X u$ is the derivative at $t = 0$ of $\Phi(t)^* u$ (see [4]).

4. F-linear derivations

Let E be an F module and let A be an F-linear transformation of E into itself. (If E is reflexive the set of F-linear transformations of E into itself can be identified with E_1^1, by §1.6). Define φ_A on (F,E) by

$$\varphi_A f = 0, \qquad f \in F,$$
$$\varphi_A X = AX, \qquad X \in E.$$

Then φ_A is clearly a derivation of (F,E), and every derivation of $(F,E$ which is F-linear and 0 on F is of this type. The induced derivations are also denoted φ_A. By (3), φ_A on E' is $-A'$, where A' is the dual of A. By (4) we have for u in E_r^s,

$$(9) \qquad (\varphi_A u)(\omega^1,\ldots,\omega^r,X_1,\ldots,X_s) = \sum_{\mu=1}^{r} u(\omega^1,\ldots,A'\omega^\mu,\ldots,\omega^r,X_1,\ldots,X_s) - \sum_{\mu=1}^{s} u(\omega^1,\ldots,\omega^r,X_1,\ldots,AX_\mu,\ldots,X_s).$$

We shall have occasion later (§7) to use a related notion.

Theorem 5. *Let E be a totally reflexive F module. There is a unique F-linear mapping Φ of E_2^2 into the F module of all F-linear mappings of the mixed tensor algebra E_*^* into itself such that for all B and C in E_1^1,*

$$(10) \qquad \Phi_{B\otimes C} = \varphi_B \cdot \varphi_C .$$

For each A in E_2^2, Φ_A maps each E_r^s into itself and each A^r into itself, and Φ_A is 0 on F. If α is in A^r then

(11) $$(\Phi_A\alpha)_{i_1\ldots i_r} = \sum_{\nu=1}^{r}(-1)^{\nu} A^{ba}_{i_\nu b}\alpha_{ai_1\ldots\hat{i}_\nu\ldots i_r} + 2\sum_{\nu<\mu}(-1)^{\mu+\nu} A^{ba}_{i_\mu i_\nu}\alpha_{abi_1\ldots\hat{i}_\nu\ldots\hat{i}_\mu\ldots i_r}\,.$$

Proof. Since E is totally reflexive, $E^2_2 = E^1_1 \otimes E^1_1$. The mapping

$$E^1_1 \times E^1_1 \longrightarrow \mathrm{Hom}_F(E^*_*, E^*_*)$$

which sends (B, C) to $\varphi_B \circ \varphi_C$ is F-bilinear and so (by the definition of tensor product) has a unique F-linear extension Φ to $E^1_1 \otimes E^1_1$. Since φ_B for B in E^1_1 sends each E^s_r and A^r into itself and is 0 on F, Φ_A has the same properties.

The notation in (11) is that of §1.12. To prove (11) it suffices to consider the case $A = B \otimes C$. By (9),

(12) $$(\varphi_C\alpha)_{i_1\ldots i_r} = \sum_{\mu=1}^{r}(-1)^{\mu} C^a_{i_\mu}\alpha_{ai_1\ldots\hat{i}_\mu\ldots i_r}\,, \qquad C \in E^1_1,\quad \alpha \in A^r,$$

where we have made use of the fact that α is alternating. If we use (12) to compute $\varphi_B\varphi_C\alpha$, we find (11) for the case $A = B \otimes C$. QED.

5. Derivations on modules which are free of finite type

Theorem 6. *Let E be an F module, free of finite type with basis $X_1, \ldots, X_n$. Let φ be a derivation of (F, E) and define γ^j_i by*

$$\varphi(X_i) = \Sigma\gamma^j_i X_j\,.$$

If $u \in E^s_r$ the components of φu are

$$(\varphi u)^{i_1\cdots i_r}_{j_1\cdots j_s} = \varphi(u^{i_1\cdots i_r}_{j_1\cdots j_s})$$

$$+ \sum_{\mu=1}^{r} \Sigma u^{i_1\cdots i_{\mu-1}a i_{\mu+1}\cdots i_r}_{j_1\cdots j_s}\gamma^{i_\mu}_a$$

$$- \sum_{\mu=1}^{s} \Sigma\gamma^a_{j_\mu} u^{i_1\cdots i_r}_{j_1\cdots j_{\mu-1}a j_{\mu+1}\cdots j_r} .$$

<u>If $X_{1'},\ldots,X_{n'}$ is another basis and $J^{j'}_i$ and $J^k_{j'}$ are defined as in §1.11, and if $\gamma^{j'}_{i'}$ is defined by</u>

$$\varphi(X_{i'}) = \Sigma\gamma^{j'}_{i'}X_{j'} ,$$

<u>then</u>

$$\gamma^{j'}_{i'} = \Sigma J^k_{i'}\gamma^{\ell}_k J^{j'}_{\ell} + \Sigma\varphi(J^k_{i'})J^{j'}_k .$$

<u>Let E be a coordinate Lie module with coordinates $x^1,\ldots,x^n$ and let the vector field X have components X^i. Then the γ^j_i corresponding to the Lie derivative θ_X are</u>

$$\gamma^j_i = -\frac{\partial X^j}{\partial x^i} .$$

<u>If X has components $X^{i'}$ with respect to new coordinates $x^{1'},\ldots,x^{n}$ then</u>

$$\frac{\partial X^{j'}}{\partial x^{i'}} = \Sigma\frac{\partial x^k}{\partial x^{i'}}\frac{\partial X^{\ell}}{\partial x^k}\frac{\partial x^{j'}}{\partial x^{\ell}} + \Sigma X^{\ell}\frac{\partial^2 x^k}{\partial x^{\ell}\partial x^{i'}}\frac{\partial x^{j'}}{\partial x^k} \tag{13}$$

<u>where we use the notation $\frac{\partial^2}{\partial x^{\ell}\partial x^{i'}}$ to mean $\frac{\partial}{\partial x^{\ell}}\frac{\partial}{\partial x^{i'}}$</u>.

<u>Proof.</u> The proof is trivial. QED.

Formula (13) shows the basic fact that the partial derivatives of the components of a tensor do not in general form the components of a tensor. This was what led Christoffel to the notion of covariant differentiation (§5).

Reference

[4] R. S. Palais, A definition of the exterior derivative in terms of Lie derivatives, Proceedings of the American Mathematical Society 5(1954), 902-908. In the definition on p.908, Φ should be identified with $(k+1)d\Theta$ rather than with $d\Theta$ itself.

1. The exterior derivative in local coordinates

Let E be an F module, A^* the Grassmann algebra (§1.9). By an exterior derivative we mean an F^0-linear mapping d of A^* into itself such that

[D1] $\quad d\colon A^r \longrightarrow A^{r+1}$,

[D2] $\quad d(\alpha\wedge\beta) = d\alpha\wedge\beta + (-1)^r\alpha\wedge d\beta$, $\qquad \alpha \in A^r$, $\beta \in A^*$,

[D3] $\quad d^2 = 0$.

Theorem 1. *Let E be a coordinate Lie module. Then there is a unique exterior derivative d such that for all scalars f, df is the differential of f.*

Proof. Let $x^1,\ldots,x^n$ be coordinates. Then each α in A^r is uniquely of the form

$$\alpha = \sum_{i_1<\cdots<i_r} \alpha_{i_1\cdots i_r} dx^{i_1}\wedge\cdots\wedge dx^{i_r} .$$

If [D2] and [D3] hold then we must have

$$d\alpha = \sum_{i_1<\cdots<i_r} \sum \frac{\partial}{\partial x^i}\alpha_{i_1\cdots i_r} dx^i\wedge dx^{i_1}\wedge\cdots\wedge dx^{i_r} .$$

That is, the components of $d\alpha$ are

$$(1) \qquad (d\alpha)_{i_1\cdots i_{r+1}} = \sum_{\mu=1}^{r+1} (-1)^{\mu+1} \frac{\partial}{\partial x^\mu} \alpha_{i_1\cdots \hat{i}_\mu \cdots i_{r+1}} .$$

This proves uniqueness.

To prove existence, choose coordinates and define d by (1), extending to all of A^* by additivity. Then d is F^0-linear, and [D1] holds. The relation [D3] holds since the $\partial/\partial x^\mu$ and ∂/∂^ν commute. To prove [D2], let $\alpha \in A^r$, $\beta \in A^s$. By the explicit formula in §1.11

for the components of $\alpha\wedge\beta$ and by (1), [D2] holds. QED.

The proof shows, of course, that (1) holds for any choice of coordinates. This is certainly the quickest approach to the exterior derivative on a manifold, for once d is known locally it is trivial to define it globally. However, a coordinate-free treatment of the exterior derivative is worthwhile for several reasons. For one, it applies to Lie modules which do not have coordinates (even locally, such as a Lie algebra over $F = F^0$). The invariant expressions for d are useful. Finally, it deepens one's understanding of the exterior derivative and shows it to be the natural dual object to the Lie product.

2. The exterior derivative considered globally

Theorem 2. *If* E *is a totally reflexive Lie module there is a unique exterior derivative* d *such that for all scalars* f , df *is the differential of* f *and for all 1-forms* ω *and vector fields* X *and* Y ,

$$2d\omega(X,Y) = X\cdot\omega(Y) - Y\cdot\omega(X) - \omega([X,Y]) . \tag{2}$$

If E *is any Lie module and we define* d *on* A^* *by*

$$(r+1)d\alpha(X_1,\ldots,X_{r+1}) = \sum_{\mu=1}^{r+1} (-1)^{\mu+1} X_\mu\cdot\alpha(X_1,\ldots,\hat{X}_\mu,\ldots,X_{r+1}) + \sum_{\mu<\nu} (-1)^{\mu+\nu}\alpha([X_\mu,X_\nu],X_1,\ldots,\hat{X}_\mu,\ldots,\hat{X}_\nu,\ldots,X_{r+1}) , \quad \alpha\in A^r, \tag{3}$$

and by extending d *to all of* A^* *by additivity, then* d *is an exterior derivative.*

Proof. For E totally reflexive, A^0 and A^1 generate A^* as an F^0 algebra, so the uniqueness assertion is clear. (Notice that the requirements that df be the differential of f and (2) are the special cases $r=0$ and $r=1$ of (3).) Therefore we need only prove that d defined by (3) is an exterior derivative.

Let $\hat{E}^r$ denote the set of all F^0-multilinear (not necessarily F multilinear) mappings of $E \times \ldots \times E$ (r times) into F. Thus $E^r \subset \hat{E}^r$. If $Z = (Z_1, \ldots, Z_r)$ is in $E \times \ldots \times E$ (r times), X is in E, and α is in $\hat{E}^r$, we use the notation $\alpha([X,Z])$ as an abbreviation for

$$\alpha([X,Z_1],Z_2,\ldots,Z_r) + \alpha(Z_1,[X,Z_2],\ldots,Z_r) + \ldots + \alpha(Z_1,Z_2,\ldots,[X,Z_r]) .$$

For α in $\hat{E}^r$ we define

$$(\partial\alpha)(X,Z) = X\cdot\alpha(Z) - \frac{1}{2}\alpha([X,Z])$$

and $(\partial\alpha)(X) = X\cdot\alpha$ if $r=0$. Then $\partial\alpha$ is in $\hat{E}^{r+1}$ and is not in general in E^{r+1} even if α is in E^r. Let $\hat{A}^r$ be the set of all alternating elements in $\hat{E}^r$, so that $A^r \subset \hat{A}^r \subset \hat{E}^r$. For α in $\hat{A}^r$ we define $d\alpha = \text{Alt}\,\partial\alpha$; that is,

$$(r+1)d\alpha(X_1,\ldots,X_{r+1}) = \sum_{\mu=1}^{r+1} (-1)^{\mu+1}\partial\alpha(X_\mu,X_1,\ldots,\hat{X}_\mu,\ldots,X_{r+1}) .$$

For α in A^r this definition of $d\alpha$ agrees with (3). It is trivial that $d\colon \hat{A}^r \longrightarrow \hat{A}^{r+1}$, and a simple computation shows that

$$d(\alpha\wedge\beta) = d\alpha\wedge\beta + (-1)^r\alpha\wedge d\beta , \qquad \alpha \in \hat{A}^r .$$

We claim that $d\colon A^r \longrightarrow A^{r+1}$; that is, if α is F-multilinear so is $d\alpha$. To see this, let $\alpha \in A^r$, $f \in F$, and let

$$\beta = (r+1)d\alpha(fX_1, X_2, \ldots, X_{r+1}) ,$$

$$\gamma = f(r+1)d\alpha(X_1, X_2, \ldots, X_{r+1}) ,$$

$$\delta = \sum_{\mu=1}^{r+1} (-1)^{\mu+1}(X_\mu \cdot f)\alpha(X_1, \ldots, \hat{X}_\mu, \ldots, X_{r+1}) .$$

We must show that $\beta = \gamma$. But

$$\beta = fX_1 \cdot \alpha(X_2, \ldots, X_{r+1}) - \frac{1}{2} \sum_{\mu=2}^{r+1} \alpha(X_2, \ldots, [fX_1, X_\mu], \ldots, X_{r+1})$$

$$+ \sum_{\mu=2}^{r+1} (-1)^{\mu+1}\{X_\mu \cdot \alpha(fX_1, \ldots, \hat{X}_\mu, \ldots, X_{r+1}) - \frac{1}{2} \alpha([X_\mu, fX_1], \ldots, \hat{X}_\mu, \ldots, X_{r+1})$$

$$- \frac{1}{2} \sum_{\nu \neq 1, \mu} \alpha(fX_1, \ldots, [X_\mu, X_\nu], \ldots, X_{r+1})\}$$

$$= \gamma - \frac{1}{2} \delta + \delta - \frac{1}{2} \delta = \gamma ,$$

which proves the claim.

It remains only to show that $d^2 = 0$. A simple computation shows that for α in $\hat{A}^r$,

$$(4) \qquad (r+2)(r+1)d^2\alpha(X_1, \ldots, X_{r+2}) =$$

$$\sum_{\mu < \nu} (-1)^{\mu+\nu}\{\partial^2\alpha(X_\mu, X_\nu, X_1, \ldots, \hat{X}_\mu, \ldots, \hat{X}_\nu, \ldots, X_{r+2})$$

$$- \partial^2\alpha(X_\nu, X_\mu, X_1, \ldots, \hat{X}_\mu, \ldots, \hat{X}_\nu, \ldots, X_{r+2})\} .$$

Let us therefore compute

$$\partial^2\alpha(X, Y, Z) - \partial^2\alpha(Y, X, Z)$$

for X and Y in E and Z in $E \times \ldots \times E$ (r times). Let $\beta(Y, Z) = \partial\alpha(Y, Z)$, so that $\partial^2\alpha(X, Y, Z) = \partial\beta(X, Y, Z)$. We have

$$\beta(Y, Z) = Y \cdot \alpha(Z) - \frac{1}{2} \alpha([Y, Z]) ,$$

so that

$$\partial^2\alpha(X,Y,Z) = X\cdot Y\cdot\alpha(Z) - \tfrac{1}{2}[X,Y]\cdot\alpha(Z) - \tfrac{1}{2}\, Y\cdot\alpha([X,Z]) - \tfrac{1}{2}\, X\cdot\alpha([Y,Z])$$
$$+ \tfrac{1}{4}\,\alpha([[X,Y],Z]) + \tfrac{1}{4}\,\alpha([Y,[X,Z]]) \ .$$

Since $X\cdot Y\cdot\alpha(Z) - Y\cdot X\cdot\alpha(Z) = [X,Y]\cdot\alpha(Z)$ we find that

$$\partial^2\alpha(X,Y,Z) - \partial^2\alpha(Y,X,Z) =$$
$$\tfrac{1}{4}\,\alpha([[X,Y],Z]) + \tfrac{1}{4}\,\alpha([Y,[X,Z]]) - \tfrac{1}{4}\,\alpha([[Y,X],Z]) - \tfrac{1}{4}\,\alpha([X,[Y,Z]])$$
$$= \tfrac{1}{4}\,\alpha([[X,Y],Z]) \ ,$$

where we have used the Jacobi identity to cancel the first, second, and fourth terms of the second line and the antisymmetry of the Lie product to re-write the third term. Consequently (4) is equal to

$$\frac{1}{4} \sum_{\mu<\nu} (-1)^{\mu+\nu} \sum_{\lambda\neq\mu,\nu} \alpha(X_1,\ldots,\hat{X}_\mu,\ldots,\hat{X}_\nu,\ldots,[[X_\mu,X_\nu],X_\lambda],\ldots,X_{r+2}) =$$

$$\frac{1}{4} \sum_{\substack{\mu<\nu \\ \lambda\neq\mu,\nu}} (-1)^{\mu+\nu}\delta^{\mu\nu\lambda}\alpha([[X_\mu,X_\nu],X_\lambda],X_1,\ldots,\hat{X}_\mu,\ldots,\hat{X}_\nu,\ldots,\hat{X}_\lambda,\ldots,X_{r+2}) =$$

$$\frac{1}{8} \sum_{\mu,\nu,\lambda} \varepsilon^{\mu\nu\lambda}\alpha([[X_\mu,X_\nu],X_\lambda],X_1,\ldots,\hat{X}_\mu,\ldots,\hat{X}_\nu,\ldots,\hat{X}_\lambda,\ldots,X_{r+2}) \ ,$$

$$\delta^{\mu\nu\lambda} = \begin{cases} (-1)^{\lambda-1} \ , & \lambda<\mu<\nu \\ (-1)^{\lambda-2} \ , & \mu<\lambda<\nu \\ (-1)^{\lambda-3} \ , & \mu<\nu<\lambda \ , \end{cases}$$

and $\varepsilon^{\mu\nu\lambda} = (-1)^{\mu+\nu+\lambda+1}$ if μ,ν,λ are in their natural order, $\varepsilon^{\mu\nu\lambda}$ changes sign under transpositions, and $\varepsilon^{\mu\nu\lambda} = 0$ if μ,ν,λ are not distinct. By the Jacobi identity again, the last sum is 0, so that $d^2 = 0$. QED.

If E is any Lie module we shall call the operator d defined by (3) the exterior derivative.

Theorem 3. *Let E be a totally reflexive F module, d an exterior derivative on A^*. Define $X\cdot f$ for X a vector field and f a scalar by*

$$(5) \qquad df(X) = X\cdot f$$

and define $[X,Y]$ for X and Y vector fields by

$$(6) \qquad 2d\omega(X,Y) = X\cdot\omega(Y) - Y\cdot\omega(X) - \omega([X,Y]) , \qquad \omega \in E' .$$

Then with respect to these operations E is a Lie module. Thus for E a totally reflexive F module there is a one-to-one correspondence between structures of Lie module on E and exterior derivatives on the Grassmann algebra.

Proof. Recall the definition of Lie module in §2.2.

Since df is a 1-form, $X \longrightarrow X\cdot$ is F-linear. Since $d(fg) = df_\wedge g + f_\wedge dg$, each $X\cdot$ is a derivation of F. Next we need to show that $[X,Y]$ is in E; that is, that $\omega([X,Y])$ is F-linear in ω. But

$$(f\omega)([X,Y]) = X\cdot((f\omega)(Y)) - Y\cdot((f\omega)(X)) - 2(d(f\omega))(X,Y)$$
$$= fX\cdot\omega(Y) + (X\cdot f)\omega(Y) - fY\cdot\omega(X) - (Y\cdot f)\omega(X) - 2(df_\wedge\omega)(X,Y) - f2d\omega(X,Y) ,$$

and by definition of the exterior product, $2(df_\wedge\omega)(X,Y) = df(X)\omega(Y) + \omega(X)df(Y) = (X\cdot f)\omega(Y) + (Y\cdot f)\omega(Y)$. Thus $(f\omega)([X,Y]) = f(\omega([X,Y]))$, and $[X,Y]$ is in E. Also,

$$0 = 2d^2f(X,Y) = X\cdot df(Y) - Y\cdot df(X) - df([X,Y])$$
$$= X\cdot Y\cdot f - Y\cdot X\cdot f - [X,Y]\cdot f .$$

Since $2d\omega(Y,X) = -2d\omega(X,Y)$, we have $[Y,X] = -[X,Y]$. Since $d\omega$ is a 2-form,

$$f2d\omega(X,Y) = 2d\omega(X,fY)$$
$$= X\cdot\omega(fY) - fY\cdot\omega(X) - \omega([X,fY])$$

so that $[X,fY] = f[X,Y] + (X\cdot f)Y$. It is clear that $[X,Y]$ is F^0-bilinear in X and Y, so it remains only to prove the Jacobi identity.

Define $\hat{d}$ by formula (3), so that $\hat{d}$ equals d on scalars and 1-forms. The proof given in Theorem 2 that $\hat{d}$ maps A^r into A^{r+1} and that $\hat{d}$ is an antiderivation did not use the Jacobi identity and so remains valid under our present assumptions. Since E is totally reflexive, A^* is generated as an F^0 algebra by A^0 and A^1. Consequently, $\hat{d} = d$. Therefore we may use (3) to compute $0 = d^2\omega(X,Y,Z)$ for ω a 1-form. Let us use $\mathfrak{S}$ to denote cyclic sums,

(7) $$\mathfrak{S}\, K(X,Y,Z) = K(X,Y,Z) + K(Y,Z,X) + K(Z,X,Y),$$

where K is any function from $E\times E\times E$ to an additive group. If α is a 2-form, (3) may be written

(8) $$3d\alpha(X,Y,Z) = \mathfrak{S}\,\{X\cdot\alpha(Y,Z) - \alpha([X,Y],Z)\}.$$

If we let $\alpha = 2d\omega$, where ω is a 1-form, we obtain

$$0 = 6d^2\omega(X,Y,Z)$$
$$= \mathfrak{S}\,\{X\cdot Y\cdot\omega(Z) - X\cdot Z\cdot\omega(Y) - X\cdot\omega([Y,Z])$$
$$-[X,Y]\cdot\omega(Z) + Z\cdot\omega([X,Y]) + \omega([[X,Y],Z])\}$$
$$= \mathfrak{S}\,\omega([[X,Y],Z]).$$

Since this is true for all 1-forms ω, the Jacobi identity holds. QED.

More generally, the proof shows that if E is an arbitrary F module, if d is an exterior derivative on A^*, and we define $X \cdot f$ for X in $E_1 = E''$ and f in F by (5) (i.e., (3) for $r=0$) and $[X,Y]$ for X and Y in E_1 by (6) (i.e., (3) for $r=1$), and if (3) holds, then E_1 is a Lie module. Thus Lie products and exterior derivatives are dual notions, and the Jacobi identity corresponds to the fact that an exterior derivative has square 0.

3. The exterior derivative and interior multiplication

Theorem 4. *Let E be a Lie module, X in E. Then the anticommutator of the exterior derivative and interior multiplication by X is the Lie derivative θ_X on exterior forms. That is,*

$$\theta_X \alpha = d(X \rfloor \alpha) + X \rfloor d\alpha , \qquad \alpha \in A^* . \tag{9}$$

If α is a closed exterior form, $\theta_X \alpha$ is exact.

Proof. We give the proof first under the assumption that E is totally reflexive, since this is the case of interest in differential geometry and the proof is less computational. Since d is an antiderivation of A^* which is homogeneous of degree 1 and interior multiplication by X is an antiderivation of A^* which is homogeneous of degree -1 (§1.10), their anticommutator is a derivation of A^* which is homogeneous of degree 0 (Theorem 1, §3.1). If E is totally reflexive, A^* is generated as an F^0 algebra by A^0 and A^1, so we need only verify (9) for α a scalar or 1-form. If $\alpha = f$ is a scalar, (9) says that $X \cdot f = df(X)$, which is true. If $\alpha = \omega$ is a 1-form, (9) says that for all vector fields Y,

$$< \theta_X \omega, Y > = < d(\omega(X)), Y > + 2d\omega(X,Y) ,$$

which is the same as (6). Thus (9) is true if E is totally reflexive. The proof for the general case is similar: one verifies (9) for α an r-form by the formula (3) for d, the definition (formula (8), §1.10) of interior multiplication, and the formula (§3.3) for Lie derivatives. The computation is omitted.

The last assertion in the theorem is an immediate consequence of (9). QED.

4. The cohomology ring

Let d be an exterior derivative on the Grassmann algebra A^* of an F module E. An exterior form α is called closed if $d\alpha = 0$, exact if $\alpha = d\beta$ for some exterior form β.

Theorem 5. *Let E be an F module, d an exterior derivative on the Grassmann algebra A^*. The set of closed exterior forms is a graded F^0 algebra in which the set of exact exterior forms is a homogeneous ideal, so that the quotient is a graded F^0 algebra.*

Proof. The proof is trivial. QED.

The quotient algebra is denoted H^*, with homogeneous subspaces H^r. It is called the cohomology ring. The dimension of H^r as an F^0 vector space is denoted b^r and called the r-th Betti number. De Rham's theorem asserts that the cohomology ring formed from the C^∞ exterior forms on a C^∞ manifold is a topological invariant, the cohomology ring of the manifold with real coefficients.

1. Affine connections in the sense of Koszul

On a differentiable manifold there is no intrinsic way of differentiating tensor fields to obtain tensor fields, covariant of one rank higher, and to obtain such a "covariant derivative" we must impose additional structure. In Riemannian geometry there is a natural notion of covariant differentiation (§7) discovered by Christoffel. Many years later Levi-Civita discovered the geometrical meaning of covariant differentiation by integrating to obtain parallel translation along curves. A number of people, especially Elie Cartan, studied non-Riemannian "affine connections," and the notion was axiomatized in a convenient way by Koszul, as follows.

Definition. Let E be a Lie module. An affine connection ∇ on E is a function $X \longrightarrow \nabla_X$ from E to the set of mappings of E into itself satisfying

$$[\nabla 0] \qquad \nabla_X(Y+Z) = \nabla_X Y + \nabla_X Z ,$$

$$[\nabla 1] \qquad \nabla_{fX+gY} = f\nabla_X + g\nabla_Y ,$$

$$[\nabla 2] \qquad \nabla_X(fY) = f\nabla_X Y + (X\cdot f)Y$$

for all vector fields X,Y,Z and scalars f and g.

Thus an affine connection is an F-linear mapping $X \longrightarrow \nabla_X$ of E into derivations on (F,E) such that for all scalars f, $\nabla_X f = X\cdot f$ (see §3.2). The derivation of the mixed tensor algebra E_*^* induced by ∇_X (§3.2) is also denoted ∇_X, and is called the covariant derivative in the direction of X (with respect to the given affine connection). In particular we have

$$(1) \qquad X\cdot\langle\omega,Y\rangle = \langle\nabla_X\omega,Y\rangle + \langle\omega,\nabla_X Y\rangle , \qquad \omega \in E', \quad Y \in E .$$

2. The covariant derivative

Covariant derivatives have an important property which is not enjoyed by Lie derivatives:

Definition. Let ∇ be an affine connection on the Lie module E, and let u be in E^s_r. We define the *covariant derivative* ∇u by

$$(2)\qquad (\nabla u)(\omega^1,\ldots,\omega^r,X_1,\ldots,X_{s+1}) = (\nabla_{X_1}u)(\omega^1,\ldots,\omega^r,X_2,\ldots,X_{s+1}) .$$

Theorem 1. Let ∇ *be an affine connection on the Lie module* E, *and let* u *be in* E^s_r. *Then* ∇u *is in* E^{s+1}_r.

Proof. Since $\nabla_{X_1}u$ is a tensor the only point at issue is the F-linearity of (2) in X_1, and this is an immediate consequence of [∇1] (which remains true for the induced derivation of the mixed tensor algebra). QED.

3. Components of affine connections

Theorem 2. Let E *be a free Lie module of finite type, with basis* $X_1,\ldots,X_n$, *and let* ∇ *be an affine connection on* E. *Define* Γ^k_{ij} *by*

$$(3)\qquad \nabla_{X_i}X_j = \Sigma\Gamma^k_{ij}X_k .$$

If u *is in* E^s_r *the components of* ∇u *are*

$$(4)\qquad (\nabla u)^{i_1\ldots i_r}_{ij_1\ldots j_s} = X_i\cdot u^{i_1\ldots i_n}_{j_1\ldots j_s}$$

$$+ \sum_{\mu=1}^{r} u^{i_1\ldots i_{\mu-1}ai_{\mu+1}\ldots i_r}_{j_1\ldots j_s}\Gamma^{i_\mu}_{ia} - \sum_{\mu=1}^{s}\Gamma^a_{ij_\mu}u^{i_1\ldots i_r}_{j_1\ldots j_{\mu-1}aj_{\mu+1}\ldots j_s} .$$

If the $\Gamma^{k'}_{i'j'}$ are defined in the same way with respect to another basis $X_{1'}, \ldots, X_{n'}$ and $J^{j'}_{i}$ and $J^{k}_{j'}$ are defined as in §1.11 then

$$(5) \qquad \Gamma^{k'}_{i'j'} = \Sigma J^{a}_{i'} J^{b}_{j'} \Gamma^{c}_{ab} J^{k'}_{c} + \Sigma (X_{i'} \cdot J^{c}_{j'}) J^{k'}_{c} .$$

Proof. This follows easily from §3.5. QED.

Notice that if the Γ^{k}_{ij} are an arbitrary set of n^3 scalars there is a unique affine connection on E satisfying (3) for the given basis. The Γ^{k}_{ij} are called the components of the affine connection (with respect to the given basis). The components may be 0 with respect to one basis but not with respect to another.

If E is a coordinate Lie module (§2.3) with coordinates $x^1, \ldots, x^n$ then (5) takes the more familiar form

$$(6) \qquad \Gamma^{k'}_{i'j'} = \Sigma \frac{\partial x^a}{\partial x^{i'}} \frac{\partial x^b}{\partial x^{j'}} \Gamma^{c}_{ab} \frac{\partial x^{k'}}{\partial x^c} + \Sigma \frac{\partial^2 x^c}{\partial x^{i'} \partial x^{j'}} \frac{\partial x^{k'}}{\partial x^c} .$$

For some applications in differential geometry, a coordinate system does not give the most convenient local basis for the vector fields. For example, on a Lie group it is usually convenient to choose a basis of left-invariant vector fields. These do not in general come from a coordinate system since they do not in general commute.

It is customary to denote the left hand side of (4) by

$$(7) \qquad \nabla_i u^{i_1 \ldots i_r}_{j_1 \ldots j_s} .$$

The notations

$$u^{i_1 \ldots i_r}_{j_1 \ldots j_s, i} \quad \text{and} \quad u^{i_1 \ldots i_r}_{j_1 \ldots j_s (i)}$$

are also in use to mean the same thing. However, it is convenient in connection with the exterior derivative (see paragraph 7) to have the new covariant index be the first covariant index.

4. Classical tensor notation for the covariant derivative.

Let ∇ be an affine connection on the Lie module E and recall the global meaning of the classical tensor notation (§1.12). Following convention, we write

$$\nabla_i u^{i_1 \cdots i_r}_{j_1 \cdots j_s} = (\nabla u)^{i_1 \cdots i_r}_{i j_1 \cdots j_s} .$$

Notice therefore that

$$\nabla_i \nabla_j u^{i_1 \cdots i_r}_{j_1 \cdots j_r} = (\nabla\nabla u)^{i_1 \cdots i_r}_{i j j_1 \cdots j_s} .$$

On the other hand, if X and Y are vector fields and u is in E^s_r, $\nabla_X \nabla_Y u$ is again in E^s_r and is in general quite different from the tensor in E^s_r obtained by substituting X and Y in the first two contravariant arguments of $\nabla\nabla u$. See paragraph 11.

5. Affine connections and tensors

Let E be a totally reflexive Lie module. We have seen (§1.6) that tensors B in E^2_1 may be regarded as F-bilinear mappings of $E \times E$ into E, and we write $B(X,Y)$ or $B_X Y$ for the image vector field. Let $\hat{E}^2_1$ be the set of all F^o-bilinear (not necessarily F-bilinear) mappings of $E \times E$ into E. Thus $\hat{E}^2_1$ is an F^o vector space, and E^2_1 is an F^o vector subspace of $\hat{E}^2_1$. Other examples of elements of $\hat{E}^2_1$ are affine connections and the Lie product. The Lie module E is called a <u>Lie algebra over</u> F in case the Lie product is in E^2_1; that is, in case $[X,Y]$ is F-bilinear in X and Y. When we say that an element of $\hat{E}^2_1$ is a tensor we mean that it lies in the vector subspace E^2_1.

Theorem 3. Let E be a totally reflexive Lie module which is not a Lie algebra over F. Then no affine connection is a tensor. The difference of any two affine connections is a tensor, and if ∇ is an affine connection and B is a tensor in E_1^2 then $\nabla + B$ is an affine connection.

Let $\nabla^{(a)}$ be a finite set of affine connections and let $\varphi^{(a)}$ be scalars. Then

$$\sum_a \varphi^{(a)} \nabla^{(a)}$$

is an affine connection if $\sum_a \varphi^{(a)} = 1$.

Proof. By [∇2] an affine connection ∇ is a tensor if and only if $(X \cdot f)Y = 0$ for all vector fields X and Y and scalars f; i.e., if and only if E is a Lie algebra over F.

Let $\nabla^{(1)}$ and $\nabla^{(2)}$ be affine connections and let $B = \nabla^{(1)} - \nabla^{(2)}$. By [∇1], $B_X Y$ is F-linear in X and by [∇2] it is F-linear in Y, so it is a tensor. If ∇ is an affine connection and B is in E_1^2 then $\nabla + B$ satisfies [∇0], [∇1], and [∇2], and so is an affine connection. The second paragraph of the theorem is obvious. QED.

By the theorem, the set of all affine connections on a totally reflexive Lie module E is either empty or is an affine subspace parallel to but disjoint from E_1^2. Thus the choice of any affine connection on E establishes a one-to-one correspondence between all affine connections and all tensor fields contravariant of rank 1 and covariant of rank 2.

Affine connections always exist on (paracompact) C^∞ manifolds.

The proof of the theorem shows that if E is totally reflexive and is a Lie algebra over F then the set of all affine connection on E is E_1^2.

6. Torsion

Definition. Let ∇ be an affine connection on the Lie module E. The torsion of ∇ is the mapping T of $E \times E$ into E defined by

$$T(X,Y) = \nabla_X Y - \nabla_Y X - [X,Y] . \tag{8}$$

The torsion tensor of ∇ is the mapping T of $E' \times E \times E$ into F defined by

$$T(\omega,X,Y) = \langle \omega, T(X,Y) \rangle . \tag{9}$$

Theorem 4. *The torsion tensor of an affine connection is a tensor.*

Proof. Since (9) is clearly F-linear in ω, we need only show that $T(X,Y)$ is F-bilinear in X and Y, and since

$$T(Y,X) = -T(X,Y) \tag{10}$$

by (8), we need only show that $T(X,Y)$ is F-linear in X. But

$$T(fX,Y) = \nabla_{fX} Y - \nabla_Y (fX) - [fX,Y]$$

$$= fT(X,Y) - (Y \cdot f)X + (Y \cdot f)X = fT(X,Y) . \qquad \text{QED.}$$

If E is totally reflexive we identify the torsion tensor and the torsion, as in the preceding paragraph (since $T \in E^2_1$). The torsion T maps E_2 into E_1 so its dual T' maps E^1 into E^2 (if E is totally reflexive), but $E^1 = A^1$ and by (10) it is clear that T' maps A^1 into A^2.

An affine connection is called torsion-free if its torsion is 0. (It is sometimes called symmetric, but this term is generally reserved for a torsion-free affine connection ∇ such that in addition

$\nabla R = 0$ where R is the curvature tensor.) The most important class of affine connections, the Riemannian connections (§7), are torsion-free.

Theorem 5. *Let* E *be a coordinate Lie module with coordinates* $x^1, \ldots, x^n$, *and let* ∇ *be an affine connection with components* Γ^k_{ij}. *Then the components of its torsion tensor* T *are*

$$T^k_{ij} = \Gamma^k_{ij} - \Gamma^k_{ji} . \tag{11}$$

Proof.

$$T^k_{ij} = T\left(dx^k, \frac{\partial}{\partial x^i}, \frac{\partial}{\partial x^j}\right) = \left< dx^k, T\left(\frac{\partial}{\partial x^i}, \frac{\partial}{\partial x^j}\right)\right>$$

$$= \left< dx^k, \nabla_{\frac{\partial}{\partial x^i}} \frac{\partial}{\partial x^j} - \nabla_{\frac{\partial}{\partial x^j}} \frac{\partial}{\partial x^i}\right> = \Gamma^k_{ij} - \Gamma^k_{ji} . \qquad \text{QED.}$$

If E is merely a free Lie module of finite type with basis $X_1, \ldots, X_n$ we define the *structure scalars* c^k_{ij} in F by

$$[X_i, X_j] = \Sigma c^k_{ij} X_k . \tag{12}$$

Then (11) must be modified to read

$$T^k_{ij} = \Gamma^k_{ij} - \Gamma^k_{ij} - c^k_{ij} .$$

Theorem 6. *Let* ∇ *be an affine connection on the Lie module* E. *Then so is* ∇' *defined by*

$$\nabla'_X Y = \nabla_Y X + [X, Y] ,$$

and $\nabla'' = \nabla$. *The affine connection*

$$\tilde{\nabla} = \frac{\nabla + \nabla'}{2} \tag{13}$$

is torsion-free. If T *is the torsion of* ∇ *then* $-T$ *is the torsion of* ∇', *and* $\nabla' = \nabla - T$.

Every affine connection ∇ on a Lie module E can be written uniquely in the form

$$\nabla = \tilde{\nabla} + \frac{1}{2}T \tag{14}$$

where $\tilde{\nabla}$ is a torsion-free affine connection and T is a skew-symmetric F-bilinear mapping of $E \times E$ into E. In this decomposition T is the torsion ∇ and $\tilde{\nabla}$ is given by (13).

Proof. Let ∇ be an affine connection with torsion T. By the definitions of T and ∇', $\nabla' = \nabla - T$, and since T is F-bilinear ∇' is an affine connection. The torsion of ∇' is given by

$$\nabla'_X Y - \nabla'_Y X - [X,Y] = \nabla_X Y - T(X,Y) - \nabla_Y X + T(Y,X) - [X,Y] = T(Y,X) = -T(X,Y).$$

By the second paragraph of Theorem 3, $\tilde{\nabla}$ defined by (13) is an affine connection, and its torsion is clearly $\frac{1}{2}$ the torsion of ∇ plus $\frac{1}{2}$ the torsion of ∇', which is 0. Thus every affine connection ∇ can be written in the form (14) where $\tilde{\nabla}$ is the torsion-free affine connection (13) and T is the torsion of ∇. It remains to prove the uniqueness of the decomposition (14), so suppose ∇ is represented in the form (14) with $\tilde{\nabla}$ a torsion-free affine connection and T a skew-symmetric F-bilinear mapping of $E \times E$ into E. Then the torsion of ∇ is the torsion of $\tilde{\nabla}$ (which is 0) plus $\frac{1}{2}T(X,Y) - \frac{1}{2}T(Y,X) = T(X,Y)$, so T is indeed the torsion of ∇. QED.

7. Torsion-free affine connections and the exterior derivative

Theorem 7. Let ∇ be a torsion-free affine connection on the Lie module E and let d be the exterior derivative. Then $d = \mathrm{Alt}\,\nabla$ on A^*.

Proof. Define $\hat{d}$ on A^* by $\hat{d} = \text{Alt}\,\nabla$. Suppose first that E is totally reflexive, so that A^* is generated as an F^0 algebra by A^0 and A^1. It is trivial to verify that $\hat{d}$ is an antiderivation of A^*, so we need only verify that $\hat{d}$ equals d on scalars and 1-forms. On scalars, ∇ is just the differential. Let ω be a 1-form. Then

$$(\nabla\omega)(X,Y) = (\nabla_X\omega)(Y) = X\cdot\langle\omega,Y\rangle - \langle\omega,\nabla_X Y\rangle$$

so that, since ∇ is torsion-free,

$$\begin{aligned} 2&(\text{Alt}\,\nabla\omega)(X,Y) \\ &= X\cdot\langle\omega,Y\rangle - \langle\omega,\nabla_X Y\rangle - Y\cdot\langle\omega,X\rangle + \langle\omega,\nabla_Y X\rangle \\ &= X\cdot\langle\omega,Y\rangle - Y\cdot\langle\omega,X\rangle - \langle\omega,[X,Y]\rangle = 2d\omega(X,Y)\,. \end{aligned}$$

If E is not totally reflexive, recall that the exterior derivative d is defined by formula (3) of §4.2. One verifies that $\hat{d} = d$ by direct computation. QED.

This is a useful theorem. The exterior derivative was defined to be Alt ∂, but ∂ is not a mapping of tensor fields into tensor fields whereas ∇ is.

8. Curvature

Definition. Let ∇ be an affine connection on the Lie module E. The curvature of ∇ is the function R from $E \times E$ into the set of mappings of E into itself given by

(15) $$R(X,Y) = \nabla_X\nabla_Y - \nabla_Y\nabla_X - \nabla_{[X,Y]}\,.$$

The <u>curvature tensor</u> of ∇ is the mapping of $E' \times E \times E \times E$ into F given by

$$R(\omega,Z,X,Y) = \langle \omega, R(X,Y)Z \rangle \ . \tag{16}$$

<u>Theorem 8.</u> <u>The curvature tensor of an affine connection is a tensor.</u>

<u>Proof.</u> Since $R(X,Y)Z$ is in E, (16) is clearly F-linear in ω. If we replace Z by fZ the coefficient is differentiated via $X\cdot Y\cdot f - Y\cdot X\cdot f - [X,Y]\cdot f = 0$, so (16) is F-linear in Z. Similarly, if we replace X by fX the coefficient is differentiated via $0 - Y\cdot f + Y\cdot f = 0$, so (16) is F-linear in X. Since

$$R(Y,X) = -R(X,Y) \tag{17}$$

it is F-linear in Y as well, and so is a tensor in E_1^3. QED.

Notice the order of the contravariant vector fields in the definition (16) of the curvature tensor. If E is totally reflexive the above identification of the curvature R with the curvature tensor is not the same as the identification $\iota(2,0,1,1)$ of E_1^3 with the set of F-linear mappings of E_2 into E_1^1 (where E_1^1 in turn is identified with the set of F-linear mappings of E into itself) but is the composition of $\iota(2,0,1,1)$ with a permutation. The definition (16) is the universally adopted convention.

<u>Theorem 9.</u> <u>Let</u> E <u>be a coordinate Lie module with coordinates</u> $x^1,\dots,x^n$ <u>and let</u> ∇ <u>be an affine connection with components</u> Γ_{ij}^k. <u>The components of the curvature tensor</u> R <u>are given by</u>

$$R_{jk\ell}^i = \frac{\partial}{\partial x^k}\Gamma_{\ell j}^i - \frac{\partial}{\partial x^\ell}\Gamma_{kj}^i + \Sigma(\Gamma_{\ell j}^a\Gamma_{ka}^i - \Gamma_{kj}^a\Gamma_{\ell a}^i) \ . \tag{18}$$

Proof.

$$R^i_{jk\ell} = \langle dx^i, (\nabla_{\frac{\partial}{\partial x^k}} \nabla_{\frac{\partial}{\partial x^\ell}} - \nabla_{\frac{\partial}{\partial x^\ell}} \nabla_{\frac{\partial}{\partial x^k}}) \frac{\partial}{\partial x^j} \rangle$$

and the computation showing that this is the same as (18) is not only trivial but easy. QED.

If E is merely a free Lie module of finite type with basis $X_1, \ldots, X_n$ we must modify (18) by replacing $\partial/\partial x^k$ by $X_k \cdot$ and $\partial/\partial x^\ell$ by $X_\ell \cdot$, and by adding the term

$$-\Sigma c^a_{k\ell} \Gamma^i_{aj}$$

where the $c^a_{k\ell}$ are the structure scalars defined by (12).

9. Affine connections on Lie algebras

Let E be a Lie algebra over $F = F^0$. As we have seen (in paragraph 5), an affine connection ∇ on E is the same as an F^0-bilinear map B of $E \times E$ into itself.

If G is a Lie group its Lie algebra $\mathfrak{g}$ (over $\mathbb{R}$) may be identified with the set of left-invariant vector fields on G . Affine connections B on $\mathfrak{g}$ such $B(X,X) = 0$ for all X in $\mathfrak{g}$ (which is the same as $B(Y,X) = -B(X,Y)$ for all X,Y in $\mathfrak{g}$) are of particular interest, for this means that each left invariant vector field X is autoparallel and the geodesics (with respect to the affine connection) issuing from the identity of G are precisely the one-parameter subgroups. The cases

$$B(X,Y) = 0 \qquad \text{(the (-) connection)},$$

$$B(X,Y) = \tfrac{1}{2}[X,Y] \qquad \text{(the (0) connection)},$$

$$B(X,Y) = [X,Y] \qquad \text{(the (+) connection)},$$

have been studied by E. Cartan and Schouten. They make sense as affine connections whenever E is a Lie algebra over F .

From the definition (8) of torsion, the torsions of the above three affine connections are respectively $-[X,Y]$, 0, and $[X,Y]$. The curvature (15) of the (-) connection is 0 since the connection itself is 0 and the curvature of the (+) connection is 0 by the Jacobi identity. For the (0) connection the curvature is given by

$$R(X,Y)Z = -\frac{1}{4}[[X,Y],Z] \ ,$$

again by the Jacobi identity. This is 0 for Abelian Lie algebras and certain nilpotent Lie algebras. Let us, as an exercise, compute ∇R . Since $R(\omega,X,Y,Z)$ is a scalar and the differential of any scalar is 0 (in the Lie algebra case) we have, for any W in E ,

$$\begin{aligned} 0 &= (\nabla_W R)(\omega,Z,X,Y) + R(\nabla_W\omega,Z,X,Y) + R(\omega,\nabla_W X,Y,Z) + R(\omega,X,\nabla_W Y,Z) \\ &\quad + R(\omega,X,Y,\nabla_W Z) = (\nabla_W R)(\omega,Z,X,Y) \\ &\quad + \frac{1}{8}\langle\omega,\{[W,[[X,Y],Z]]-[[X,Y],[W,Z]]-[[[W,X],Y],Z] - [[X,[W,Y]],Z]\}\rangle \\ &= (\nabla_W R)(\omega,Z,X,Y) \end{aligned}$$

since, by the Jacobi identity, θ_W is a derivation (cf. §2.1). Thus the (0) connection is symmetric; that is, the torsion is 0 and $\nabla R = 0$.

We shall not return to this subject. See Helgason [2,§1].

10. The Bianchi identities

Following Nomizu [5] we prove some identities relating curvature and torsion. As in §4.2, $\mathfrak{S}$ denotes cyclic sums.

<u>Theorem 10.</u> <u>Let</u> ∇ <u>be an affine connection on a Lie module</u> E <u>with torsion</u> T <u>and curvature</u> R . <u>Then, for all vector fields</u> X,Y,Z , <u>the following identities hold:</u>

(19) $$T(X,Y) = -T(Y,X) ,$$

(20) $$R(X,Y) = -R(Y,X) ,$$

(21) $$\mathfrak{S}\, R(X,Y)Z = \mathfrak{S}\, T(T(X,Y),Z) + \mathfrak{S}\, (\nabla_X T)(Y,Z) ,$$

(22) $$\mathfrak{S}\, (\nabla_Z R)(X,Y) + \mathfrak{S}\, R(T(X,Y),Z) = 0 .$$

<u>In particular, if</u> ∇ <u>is torsion-free</u>

(23) $$\mathfrak{S}\, R(X,Y)Z = 0 ,$$

(24) $$\mathfrak{S}\, (\nabla_Z R)(X,Y) = 0 .$$

<u>Proof.</u> We have already noted the trivial identities (19) and (20). To prove (21) and (22) we need

(25) $$(\nabla_Z T)(X,Y) = \nabla_Z(T(X,Y)) - T(\nabla_Z X,Y) - T(X,\nabla_Z Y) ,$$

(26) $$(\nabla_Z R)(X,Y) = [\nabla_Z, R(X,Y)] - R(\nabla_Z X,Y) - R(X,\nabla_Z Y) .$$

The relation (25) is an immediate consequence of the fact that ∇_Z is a derivation of the mixed tensor algebra. For (26) we need Theorem 3, §3.2 as well (notice that $[\nabla_Z, R(X,Y)]$ is not a Lie product of vector field but a commutator of operators on vector fields).

To prove (21), we use the definition (8) of torsion and (19) to find

(27) $$T(T(X,Y),Z) = T(\nabla_X Y,Z) + T(Z,\nabla_Y X) - T([X,Y],Z) .$$

We use (25) to re-express the first two terms on the right of (27) and make the discovery that

$$\mathfrak{S}T(T(X,Y),Z) = -\mathfrak{S}(\nabla_Z T)(X,Y) + \mathfrak{S}\nabla_Z(T(X,Y)) - \mathfrak{S}T([X,Y],Z)$$

$$= \mathfrak{S}\{-(\nabla_Z T)(X,Y) + \nabla_Z\nabla_X Y - \nabla_Z\nabla_Y X - \nabla_Z[X,Y] - \nabla_{[X,Y]}Z + \nabla_Z[X,Y] - [[X,Y],Z]\}$$

$$= \mathfrak{S}\{-(\nabla_Z T)(X,Y) + R(X,Y)Z\}$$

by the Jacobi identity and the definition (15) of curvature. Thus (21) holds.

To prove (22), use the definition (8) of torsion and (20) to find

$$R(T(X,Y),Z) = R(\nabla_X Y - \nabla_Y X - [X,Y],Z)$$

$$= R(\nabla_X Y,Z) + R(Z,\nabla_Y X) - R([X,Y],Z) .$$

Sum cyclically and use (26) to obtain

$$\mathfrak{S}R(T(X,Y),Z) = \mathfrak{S}\{-(\nabla_Z R)(X,Y) + [\nabla_Z,R(X,Y)] - R([X,Y],Z)\} .$$

By the definition (15) of curvature and the Jacobi identity,

$$\mathfrak{S}\{[\nabla_Z,R(X,Y)] - R([X,Y],Z)\}$$

$$= \mathfrak{S}\{[\nabla_Z,[\nabla_X,\nabla_Y] - \nabla_{[X,Y]}] - [\nabla_{[X,Y]},\nabla_Z] + \nabla_{[[X,Y],Z]}\}$$

$$= \mathfrak{S}\{[\nabla_Z,[\nabla_X,\nabla_Y]] + \nabla_{[[X,Y],Z]}\} = 0 .$$ QED.

The relation (24) (and sometimes (23)) for a torsion-free affine connection is called <u>Bianchi's identity.</u> If ∇ is a torsion-free affine connection with curvature tensor R then

(28) $$R^i_{jk\ell} + R^i_{k\ell j} + R^i_{\ell jk} = 0 ,$$

(29) $$\nabla_m R^i_{jk\ell} + \nabla_k R^i_{j\ell m} + \nabla_\ell R^i_{jmk} = 0 ,$$

(30) $$R^i_{jk\ell} = -R^i_{j\ell k} ,$$

since these are merely (23), (24), and (20) in a different notation.

11. Ricci's identity

The identities we prove next will be used frequently. Recall the definition of φ_A given in §3.4, where A is an F-linear mapping of an F module E into itself. If ∇ is an affine connection on a Lie module E and u is in E_r^s we will use the notations $(\nabla u)(X)$ and $(\nabla\nabla u)(X,Y)$, it being understood that X and X,Y are the first contravariant arguments in ∇u and $\nabla\nabla u$ respectively. Thus $(\nabla u)(X)$ and $(\nabla\nabla u)(X,Y)$ are again tensors in E_r^s.

Theorem 11. *Let ∇ be an affine connection on the Lie module E with torsion T and curvature R. Then for all vector fields X,Y and all mixed tensors u,*

$$(31) \qquad \nabla_X\nabla_Y u - \nabla_Y\nabla_X u - \nabla_{[X,Y]}u = \varphi_{R(X,Y)}u ,$$

$$(32) \qquad (\nabla\nabla u)(X,Y) - (\nabla\nabla u)(Y,X) = \varphi_{R(X,Y)} - \nabla_{T(X,Y)}u .$$

In particular, if ∇ is torsion-free then

$$(33) \qquad (\nabla\nabla u)(X,Y) - (\nabla\nabla u)(Y,X) = \varphi_{R(X,Y)}u$$

is F-linear in u.

Proof. For given X,Y in E, let

$$\varphi = [\nabla_X,\nabla_Y] - \nabla_{[X,Y]} .$$

By §3.2 this is the derivation on the mixed tensor algebra induced by the derivation $[\nabla_X,\nabla_Y] - \nabla_{[X,Y]}$ on E, so to prove (31) we need verify it only on scalars (where both sides are 0) and vector fields (where it is the definition of curvature).

To prove (32), let u be in E^s_r, z in $\overset{0r}{E}_s$. Then

$$(\nabla\nabla u)(X,Y,z) - (\nabla\nabla u)(Y,X,z)$$

$$= (\nabla_X \nabla u)(Y,z) - (\nabla_Y \nabla u)(X,z)$$

$$= X\cdot((\nabla u)(Y,z)) - (\nabla u)(\nabla_X Y,z) - (\nabla u)(Y,\nabla_X z)$$

$$-Y\cdot((\nabla u)(X,z)) + (\nabla u)(\nabla_Y X,z) + (\nabla u)(X,\nabla_Y z)$$

$$= (\nabla_X \nabla_Y u)(z) - (\nabla_{\nabla_X Y} u)(z) - (\nabla_Y \nabla_X u)(z) + (\nabla_{\nabla_Y X} u)(z)$$

so that, by the definition (8) of torsion,

$$(\nabla\nabla u)(X,Y) - (\nabla\nabla u)(Y,X)$$

$$= (\nabla_X \nabla_Y - \nabla_Y \nabla_X - \nabla_{[X,Y]})u - \nabla_{T(X,Y)} u .$$

By (31), this proves (32). QED.

If ∇ is torsion-free we may write (33) as

(34) $$\nabla_\ell \nabla_k u^{i_1 \cdots i_r}_{j_1 \cdots j_s} - \nabla_k \nabla_\ell u^{i_1 \cdots i_r}_{j_1 \cdots j_s}$$

$$= \sum_{\mu=1}^{s} R^a_{j_\mu k\ell} u^{i_1 \cdots i_r}_{j_1 \cdots j_{\mu-1} a j_{\mu+1} \cdots j_s}$$

$$- \sum_{\mu=1}^{r} u^{i_1 \cdots i_{\mu-1} a i_{\mu+1} \cdots i_r}_{j_1 \cdots j_s} R^{i_\mu}_{ak\ell} .$$

This is <u>Ricci's identity</u>. We emphasize again that we follow the custom of writing

$$\nabla_\ell \nabla_k u^{i_1 \cdots i_r}_{j_1 \cdots j_s} \quad \text{for} \quad (\nabla\nabla u)^{i_1 \cdots i_r}_{\ell k j_1 \cdots j_s} .$$

12. Twisting and turning

Suppose we have an affine connection on the manifold M. Let C be a parameterized curve $t \longrightarrow C_t$ in M with tangent vector X, and let Y be a vector field on M. In local coordinates X has components $dx^k(t)/dt$ at C_t, where $x^k(t)$ is the k-th coordinate of C_t. Let Y have components $y^k(t)$ at C_t. Then $\nabla_X Y$ at C_t has components

$$\frac{dy^k}{dt} + \Sigma\Gamma^k_{ij}\frac{dx^i}{dt}y^j \ . \tag{35}$$

That is, we don't need to know Y off C to compute $\nabla_X Y$ since X points in the direction of C.

If we prescribe $y^k(0)$ as the components of a tangent vector Y_p at the initial point $p = C_0$ of C and set (35) equal to 0, we have a well-posed initial value problem. When we integrate we obtain a family of vectors Y along C satisfying $\nabla_X Y = 0$ and $Y = Y_p$ at p. Let $q = C_1$ be the final point of the curve segment C $(0 \leq t \leq 1)$. Then we obtain a mapping

$$\tau_c : M_p \longrightarrow M_q$$

by setting $\tau_c Y_p = Y_q$. This mapping is linear and invertible, and we obtain a groupoid in this way. The mapping τ_c is called parallel translation along C.

A particular case of (35) is $\nabla_X X$ along C, which has components

$$\frac{d^2x^k}{dt^2} + \Sigma\Gamma^k_{ij}\frac{dx^i}{dt}\frac{dx^j}{dt} \ . \tag{35}$$

This is called the acceleration vector. Notice that although the velocity vector of a particle moving on a differentiable manifold makes sense, there is no meaning to the notion of acceleration vector unless we have additional structure such as an affine connection. Notice that (36) is the same for affine connections having the same torsion-free part (see Theorem 6), since $(dx^i/dt)(dx^j/dt)$ is symmetric in i and j. If we start a particle at p at time 0 with prescribed initial velocity X_p and required the acceleration (36) to be 0, we have a well-posed initial value problem for a second-order differential equation. This second-order differential equation has the special property that if we multiply X_p by a constant k the particle travels in the same trajectory with velocity k times the previous velocity, and is called a spray. In fact, this is the geodesic spray of the affine connection and the trajectories are called the geodesics of the affine connection. Affine connections with the same torsion-free part give rise to the same geodesic spray and the same geodesics. Any spray is the geodesic spray of a unique torsion-free affine connection. On a manifold with an affine connection (or with a spray) Newton's dynamical law $F = ma$ is meaningful. (The proper setting for Hamiltonian dynamics is a symplectic manifold (§8).)

Suppose we have a torsion-free affine connection ∇ on M. Let p be a point in M and let X_p, Y_p be two tangent vectors at p Choose curves at p with tangent vectors X_p, Y_p and let q and r be the points on these curves corresponding to the parameter value t (see Figure 4). Parallel translate Y_p to q to obtain Y_q and parallel translate X_p to r to obtain Y_r, and choose curves at r and q with these tangent vectors. Since ∇ is torsion-free and $\nabla_X Y$ and $\nabla_Y X$ are 0 at p, $[X,Y]$ is 0 at p. Therefore (§2.4) the end points (corresponding to the parameter value t) of the curves

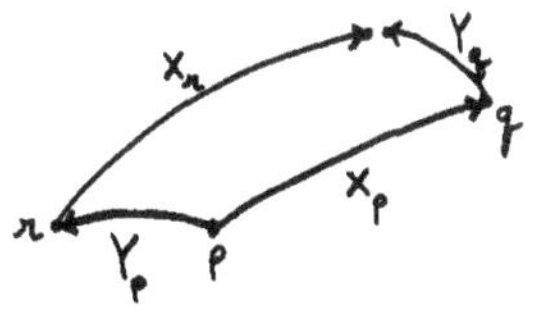

Figure 4. A parallelogram

representing X_r and Y_q are equal to second order in t . This is false if the affine connection has torsion.

Given a spray or affine connection on M , let Y_p be a tangent vector at p and let a particle start at p with velocity Y_p and travel for unit time with zero acceleration. Its position at unit time is denoted $\exp Y_p$, so that exp (the exponential mapping) is a mapping of the tangent-space M_p into the manifold M . It is always well-defined locally (and locally is a diffeomorphism) and is sometimes well-defined globally (in which case the spray or affine connection is called complete).

Let C be a curve segment in M starting at p with tangent vectors X , and given an affine connection ∇ on M let $Y_p(t)$ be the parallel translate of Y_p in M_p along C to C_t . For Y_p sufficiently small, the curves $\exp Y_p(t)$ sweep out a tubular neighborhood of C . If ∇ is torsion-free then the tangent vectors to these curves are, to first order, obtained by parallel translation of X as in Figure 4. Now consider the affine connection with the same torsion-free part but with torsion T. Then in (35) we add the term

$$\frac{1}{2}\sum T^k_{ij}\frac{dx^i}{dt}y^j .$$

Since $T^k_{ii} = 0$, $\frac{1}{2}\sum T^k_{ij}\frac{dx^i}{dt}$ is a linear transformation acting transversally to the direction of the curves $\exp Y_p(t)$, and it twists them around C. This is torsion.

Now let us consider curvature. A frame at a point p in the manifold M is simply a basis of the tangent space M_p. The set of all frames at p is the principal homogeneous space of the general linear group $GL(n,\mathbb{R})$. If any frame is singled out, there is a unique element of $GL(n,\mathbb{R})$ taking it into any given frame. The set of all frames at all points is a principal fiber bundle over M with structural group $GL(n,\mathbb{R})$. The projection maps each frame to the point p in M at which it lives.

Let ∇ be an affine connection on M, C a curve segment on M. Then parallel translation along C takes frames at the initial point p into frames at the final point q. The frame travels along C to q, picks up a tensor and brings it back to p for differentiation or other purposes. It is similar to the repair truck which leaves the Mobil station to tow back a car and in fact the notion is called the repère mobile. If C is a closed loop at p parallel translation around C gives an automorphism of M_p, and the set of such automorphisms forms a group called the holonomy group.

Recall from §2.4 the geometrical meaning of the Lie product of two vector fields X and Y, which may be expressed by saying that

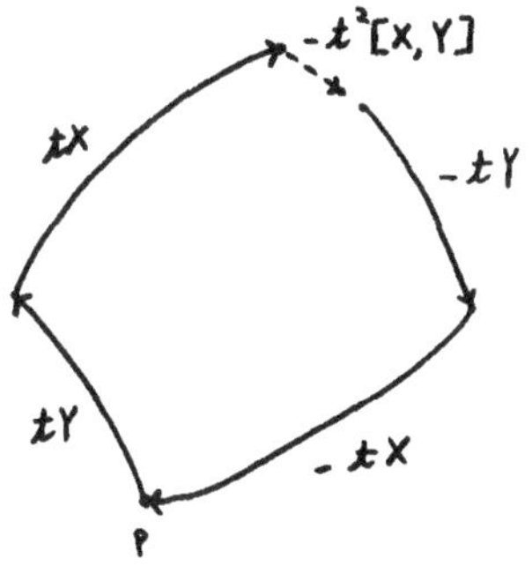

Figure 5. A loop

Figure 5 is a closed loop to second order in t. If we parallel translate a frame around this loop we will find in general that it has turned. In fact, parallel translation of a tangent vector Z around the loop gives to second order in t

$$Z + t^2(\nabla_X \nabla_Y Z - \nabla_Y \nabla_X Z - \nabla_{[X,Y]} Z) = Z + t^2 R(X,Y)Z .$$

This is curvature.

Reference

[5] K. Nomizu, Lie Groups and Differential Geometry, Publication of the Mathematical Society of Japan 2, (1956).

§6. Holonomy

1. Principal fiber bundles

The frame bundle discussed in §5.12 is an example of a principal fiber bundle. In a principal fiber bundle we have a C^∞ mapping $\pi: P \longrightarrow B$ of the bundle P onto the base B, where P and B are C^∞ manifolds. Each fiber $\pi^{-1}(p)$, $p \in B$, is a principal homogeneous space of a Lie group G, the structure group of the bundle. That is, each fiber is the same as G except that it has forgotten which element is the identity (cf. frames vs. the general linear group). The structure group G acts by translations on each fiber and therefore on the whole bundle P, and this action is C^∞.

One can define the notion of connection in this setting and discuss curvature and holonomy and the relation of connections to reductions of the structure group (see Nomizu [5,§5]). The discussion is simpler if we confine ourselves entirely to G-invariant objects on the bundle P. The invariant scalars on P are isomorphic to the algebra of all scalars on B, and the invariant vector fields on P form a Lie module over them.

Each invariant vector field on P projects onto a vector field on B, and some of them (those which lie along the fibers) project onto 0. We shall discuss this situation algebraically. Because of our restriction to G-invariant objects, the terminology is a bit different from the standard terminology [5,§5]. The usual treatment is complicated by the notion of the connection form, a notion which we avoid.

At each point in a principal fiber bundle we have a distinguished linear subspace of the tangent space, called the vertical space, consisting of those tangent vectors tangent to the fiber. A connection is

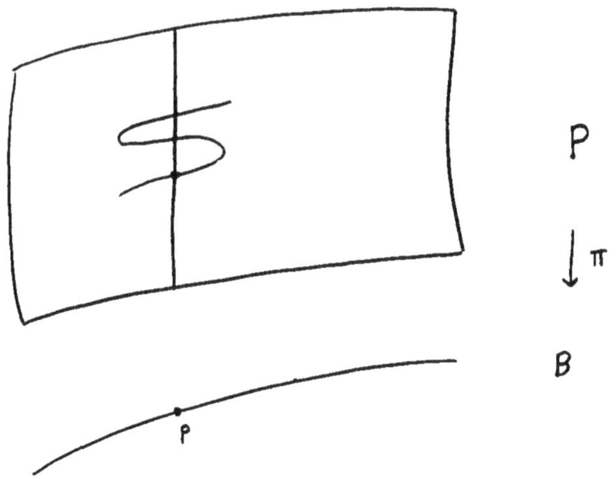

Figure 6. A horizontal curve

a G-invariant choice of a complementary linear subspace at each point, called the horizontal space. The field of horizontal subspaces is not in general integrable, and this is where the notion of holonomy enters (see §2.4). A curve in P is called horizontal if its tangent vector always lies in the horizontal space. A horizontal curve may cut a given fiber in many points (see Fig.5), and the curvature of the connection describes the vertical directions of motion which can be approximated by horizontal curves.

2. Lie bundles

The collection of all Lie modules over a fixed F forms a category if we define a morphism of two Lie modules Q and P to be an F-module homomorphism $\rho: Q \longrightarrow P$ (i.e., an F-linear mapping of Q into P such that

$$(\rho X)\cdot f = X\cdot f\,, \qquad X \in Q,\ f \in F\,,$$

$$\rho[X,Y] = [\rho X,\rho Y]\,, \qquad X,\ Y \in Q\,.$$

A morphism in this category will be called a Lie homomorphism.

An ideal V in a Lie module P is a submodule such that

$$[X,Z] \in V\,, \qquad X \in P,\ Z \in V\,,$$

and such that

$$Z\cdot f = 0\,, \qquad Z \in V,\ f \in F\,.$$

If P is a Lie module then $\{X \in P\colon X\cdot f = 0 \text{ for all } f \in F\}$ is clearly an ideal, and every ideal is contained in it.

If V is an ideal in the Lie module P then the quotient module P/V is in the obvious way a Lie module, and the projection $\pi\colon P \to P/V$ is a Lie homomorphism. If $\rho\colon P \longrightarrow Q$ is any Lie homomorphism then the kernel V of ρ is an ideal, the image $\rho(P)$ is a Lie module, and the induced mapping of P/V onto $\rho(P)$ is an isomorphism.

A submodule K of a Lie module P is called a Lie submodule in case $[X,Y] \in K$ whenever X and Y are in K. Then K is a Lie module and the inclusion $\iota\colon K \longrightarrow P$ is a Lie homomorphism.

Let P be a Lie module and H a submodule (not necessarily a Lie submodule), and let ρ be the projection of P onto the module P/H. For X and Y in H define

$$R(X,Y) = \rho[X,Y]\,.$$

We call R the curvature of the submodule.

Theorem 1. *Let H be a submodule of the Lie module P. Then the curvature R is F-bilinear and antisymmetric, and is 0 if and only if H is a Lie submodule.*

Proof. The curvature is F-linear in X since

$$R(fX,Y) = \rho[fX,Y] = f\rho[X,Y] - (Y\cdot f)\rho X = f\rho[X,Y] = fR(X,Y) .$$

Since R is clearly antisymmetric it is also F-linear in Y. The last statement in the theorem is obvious. QED.

Definition. A *Lie bundle* $\pi\colon P \to B$ is an epimorphism of Lie modules. The kernel V of π is called the *vertical module* and elements of it are called *vertical*. A *reduction* of a Lie bundle $\pi\colon P \to B$ is a commutative diagram

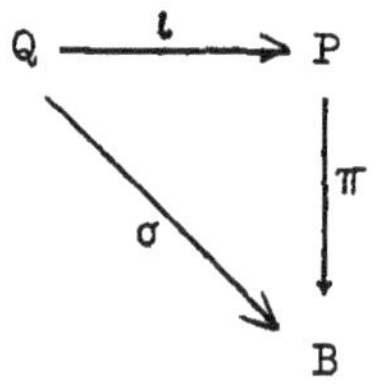

where $\iota\colon Q \to P$ is a monomorphism of Lie modules and $\sigma\colon Q \to B$ is a Lie bundle. A Lie bundle $\pi\colon P \to B$ is *trivial* if there is a reduction with $Q = B$ and σ the identity. A *connection* in a Lie bundle $\pi\colon P \to B$ is a submodule H of P such that P is the module direct sum of H and the vertical module V. The module H in a connection is also called the *horizontal module* and elements of it are called *horizontal*. The projections of P onto V and H are denoted v and h respectively. The *curvature* of a connection is the mapping R of $H \times H$ into V given by

$$R(X,Y) = v[X,Y] , \qquad X,Y \in H .$$

A connection is *flat* in case its curvature is 0. The *holonomy module* L of a connection is the smallest submodule of P such that

$$\text{(1)} \qquad R(X,Y) \in L \,, \qquad X,Y \in H \,,$$

and

$$\text{(2)} \qquad [X,Z] \in L \,, \qquad X \in H, \quad Z \in L \,.$$

Notice that we may identify the module P/H with V, so that the definition of curvature agrees with the previous one. We may if we wish extend R to a mapping of $P \times P$ into V by setting

$$R(X,Y) = v[hX,hY] \,, \qquad X,Y \in P \,.$$

This has the effect of setting $R(X,Y) = 0$ if X or Y is vertical.

The horizontal module H in a connection of the Lie bundle $\pi\colon P \longrightarrow B$ is isomorphic as a module to B, since it is complementary to the kernel V of π. However, if the connection is not flat then H is not a Lie module whereas B is. If X is in B the element $\tilde{X}$ of H such that $\pi\tilde{X} = X$ is called the *lift* of X. We also define the *curvature* as a mapping of $B \times B$ into V by setting $R(X,Y) = R(\tilde{X},\tilde{Y})$.

Theorem 2. Let $\pi\colon P \longrightarrow B$ *be a Lie bundle with vertical module* V. *Suppose there is a connection with horizontal module* H, *and let* L *be the holonomy module.*

Then L *is contained in* V, *the modules* L *and* $H+L$ *are Lie modules,* L *is an ideal in* $H+L$, *and*

$$\begin{array}{ccc} H+L & \xrightarrow{\ \iota\ } & P \\ & \searrow^{\sigma} & \downarrow{\scriptstyle \pi} \\ & & B \end{array}$$

where ι *is the inclusion map and* σ *is the restriction of* π, *is a reduction of the Lie bundle* $\pi\colon P \longrightarrow B$. *Also,* H *is a connection in*

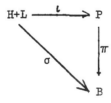

$\sigma: H+L \longrightarrow B$ <u>and its holonomy module is</u> L. <u>A Lie bundle is trivial if and only if it has a flat connection.</u>

<u>Proof.</u> By definition of the curvature, V satisfies (1) and since V is an ideal (being the kernel of π) it satisfies (2). But L is the smallest module satisfying (1) and (2), so $L \subset V$.

The holonomy module L is spanned as a module by the set of all vector fields of the form

$$(3) \qquad Z = \theta_{X_n} \cdots \theta_{X_2} R(X_1, X_0), \qquad X_i \in H, \quad 0 \leq i \leq n,$$

since each

$$\theta_{X_{n+1}}(fZ) = f\theta_{X_{n+1}} Z + (X_{n+1} \cdot f)Z$$

is an F-linear combination of elements of the form (3).

Let Z be given by (3) and let

$$(4) \qquad W = \theta_{Y_m} \cdots \theta_{Y_2} R(Y_1, Y_0), \qquad Y_i \in H, \quad 0 \leq i \leq m$$

be of the same form. By the Jacobi identity,

$$(5) \qquad [W,Z] = \theta_{Y_m} [\theta_{Y_{m-1}} \cdots \theta_{Y_2} R(Y_1, Y_0), Z] - [\theta_{Y_{m-1}} \cdots \theta_{Y_2} R(Y_1, Y_0), \theta_{Y_m} Z].$$

Assume as an induction hypothesis on m that for all W of the form (4) and all Z of the form (3) for any value of n, the Lie product $[W,Z]$ is in L. The relation (5) shows that if $m \geq 2$ the induction hypothesis is true for m if it is true for $m-1$. Consequently, to show that $[W,Z]$ is always in L we need only show that

$$[R(Y_1, Y_0), Z], \qquad Y_1, Y_0 \in H, \quad Z \in L,$$

is in L. By the definition of R,

(6) $$[Y_1,Y_0] = R(Y_1,Y_0) + h[Y_1,Y_0] .$$

But $h[Y_1,Y_0]$ is horizontal, so that $[h[Y_1,Y_0],Z]$ is in L. Therefore we need only show that

$$[[Y_1,Y_0],Z] , \qquad Y_1,Y_0 \in H, \quad Z \in L ,$$

is in L. But by the Jacobi identity

$$[[Y_1,Y_0],Z] = [Y_1,[Y_0,Z]] - [Y_0,[Y_1,Z]] ,$$

and the right hand side is in L. Therefore L is a Lie module.

Since L is contained in V, the module $H+L$ is a direct sum of modules. Let $X_1,X_2 \in H$ and $Z_1,Z_2 \in L$. Then

$$[X_1 + Z_1, X_2 + Z_2] = [X_1,X_2] + [X_1,Z_2] + [Z_1,X_2] + [Z_1,Z_2] .$$

Since L is a Lie module, $[Z_1,Z_2]$ is in L, and $[X_1,Z_2] + [Z_1,X_2]$ is in L by the definition of L. The formula (6) applied to $[X_1,X_2]$ shows that it is in $H+L$. Therefore $H+L$ is a Lie module, and the argument shows that L is an ideal in $H+L$ (if we notice in addition that $Z\cdot f = 0$ for all Z in L and scalars f since L is contained in the ideal V). Consequently, if we let σ be the restriction of π to $H+L$ then $\sigma\colon H+L \longrightarrow B$ is a Lie bundle and, together with the inclusion $\iota\colon H+L \longrightarrow P$, gives a reduction of $\pi\colon P \longrightarrow B$. It is clear that H is a connection in $\sigma\colon H+L \longrightarrow B$ and that its holonomy module is again L.

In particular, if the connection is flat then $L = 0$ since 0 satisfies (1) and (2). Then H is a Lie module, $\sigma\colon H \longrightarrow B$ is an isomorphism of Lie modules, and $\iota \circ \sigma^{-1}\colon B \longrightarrow P$ is a monomorphism of

Lie modules, so that $\pi: P \longrightarrow B$ is trivial. Conversely, if $\pi: P \rightarrow B$ is trivial we may take as our connection H the image of B in P in the trivialization, and H is a flat connection. QED.

3. The relation between the two notions of connection

On a manifold we have the notion of an affine connection in the sense of Koszul (§5.1) and the notion of a connection in the frame bundle, and the notions have a close affinity. There is a one-to-one correspondence between $GL(n, \mathbb{R})$-invariant vector fields on the frame bundle and derivations (§3.2) of the module of vector fields, and this motivates the following construction.

Theorem 3. Let B be the Lie module of all derivations of F and let P be the module of all derivations of (F,B). For φ in P define $\pi\varphi$ in B to be the restriction of φ to F. With respect to the operations $\varphi \cdot f = (\pi\varphi) \cdot f$ and $[\varphi_1, \varphi_2] = \varphi_1\varphi_2 - \varphi_2\varphi_1$, the module P is a Lie module and $\pi: P \longrightarrow B$ is a Lie bundle. The vertical module V is the set of all derivations of (F,B) which are 0 on F and F-linear on B.

Let H be a connection in $\pi: P \longrightarrow B$. Then the mapping of B into H which takes X into its lift $\tilde{X}$ is an affine connection (in the sense of Koszul). Conversely, if ∇ is an affine connection the set H of all ∇_X for X in B is a connection in $\pi: P \longrightarrow B$, and ∇_X is the lift of X. The curvature R as a mapping $B \times B$ into V is identical with the curvature of the affine connection ∇.

Proof. By Theorem 1, §2.2, B is indeed a Lie module. By §3.2 the set P of all derivations of (F,B) is an F^{o} Lie algebra and it is readily seen to be a Lie module, and $\pi\colon P \longrightarrow B$ is clearly a Lie bundle, with vertical module as described.

Let H be a connection, and define ∇_X to be the lift $\tilde{X}$ of X. Then $X \longrightarrow \nabla_X$ is F-linear and each ∇_X is a derivation of (F,B) so ∇ is an affine connection. Conversely, if ∇ is an affine connection the set H of all ∇_X is a submodule of P, since $X \longrightarrow \nabla_X$ is F-linear. We need to show that any derivation φ of (F,B) is the sum of some ∇_X and an element of V, but if we let $X = \pi\varphi$ then $\varphi - \nabla_X$ is clearly in V. Equally clearly, $V \cap H = 0$, so H is a connection. By definition ∇_X is horizontal and since $\pi\nabla_X = X$, ∇_X is the lift of X. Thus we may identify the two notions of connection.

The curvature as a mapping of $B \times B$ into V is by definition

$$R(X,Y) = R(\tilde{X},\tilde{Y}) = v[\tilde{X},\tilde{Y}] = [\tilde{X},\tilde{Y}] - h[\tilde{X},\tilde{Y}] .$$

Since $\pi[\tilde{X},\tilde{Y}] = [\pi\tilde{X},\pi\tilde{Y}] = [X,Y]$, the horizontal part of $[\tilde{X},\tilde{Y}]$ is the lift of $[X,Y]$. That is,

$$R(X,Y) = [\nabla_X,\nabla_Y] - \nabla_{[X,Y]} . \qquad \text{QED.}$$

§7. Riemannian metrics

1. Pseudo-Riemannian metrics

Let E be an F module and let u be a covariant tensor of rank 2. If X is any vector field we let $u(X)$ be the 1-form defined by

$$\langle u(X), Y\rangle = u(X,Y) .$$

The tensor u is called non-degenerate in case the mapping $u: E \to E'$ is bijective. If u is non-degenerate $u: E \to E'$ is an isomorphism of F modules, so that E is reflexive. Therefore the dual u' is also a mapping of E to E', and u is symmetric if and only if $u = u'$. The inverse mapping of $u: E \to E'$ for u non-degenerate is denoted u^{-1}, so that $u^{-1}: E' \to E$. The inverse mapping u^{-1} determines the contravariant tensor u^{-1} of rank 2 given by $u^{-1}(\omega,\eta) = \langle u^{-1}\omega, \eta\rangle$.

Definition. A pseudo-Riemannian metric on an F module E is a symmetric non-degenerate covariant tensor of rank 2.

On a manifold M, a Riemannian metric g is a pseudo-Riemannian metric such that $g(X,X) \geq 0$ for all vector fields X (and consequently $g(X,X)(p) > 0$ at all points p such that $X_p \neq 0$). Riemannian metrics enjoy important geometrical and analytic properties not shared by pseudo-Riemannian metrics, but the formal algebraic properties, which we will consider now, are the same.

If the F module E is free of finite type with basis $X_1,\dots,X_n$ and g is a pseudo-Riemannian metric we denote its components by g_{ij} and those of g^{-1} by g^{ij}, so that

$$\Sigma g_{ij} g^{jk} = \delta_i^k , \qquad \Sigma g^{ij} g_{jk} = \delta_k^i ,$$

and the g^{ij} are given explicitly in terms of the $g_{k\ell}$ by the familiar but extraordinarily complicated formula for the inverse of a matrix.

2. The Riemannian connection

Theorem 1. *Let* g *be a pseudo-Riemannian metric on a Lie module* E . *Then there is a unique torsion-free affine connection* ∇ *such that* $\nabla g = 0$.

Proof. First we prove uniqueness. Let X, Y , and Z be vector fields. Since $\nabla g = 0$ we have $\nabla_Z g = 0$, so that

$$Z \cdot g(X,Y) = g(\nabla_Z X, Y) + g(X, \nabla_Z Y) .$$

Since ∇ is torsion-free,

$$Z \cdot g(X,Y) = g(\nabla_X Z, Y) + g([Z,X],Y) + g(X, \nabla_Z Y) , \tag{1}$$

$$X \cdot g(Y,Z) = g(\nabla_Y X, Z) + g([X,Y],Z) + g(Y, \nabla_X Z) , \tag{2}$$

$$Y \cdot g(Z,X) = g(\nabla_Z Y, X) + g([Y,Z],X) + g(Z, \nabla_Y X) , \tag{3}$$

where (2) and (3) are obtained from (1) by cyclic permutation. Now subtract (2) from (1) plus (3). We obtain

$$\begin{aligned} 2g(X, \nabla_Z Y) &= Z \cdot g(X,Y) + Y \cdot g(Z,X) - X \cdot g(Y,Z) \\ &\quad -g([Z,X],Y) - g([Y,Z],X) + g([X,Y],Z) . \end{aligned} \tag{4}$$

The right hand side of (4) does not involve ∇ , so we have a formula for $g(\nabla_Z Y)$ on X . Since g is non-degenerate and X is arbitrary, $\nabla_Z Y$ is uniquely determined, so ∇ is unique.

To prove existence, consider the right hand side of (4). This is F-linear in X , for if we replace X by fX then f is differentiated via

$$(Z\cdot f)g(X,Y) + (Y\cdot f)g(Z,X) - (Z\cdot f)g(X,Y) - (Y\cdot f)g(X,Z) = 0 .$$

Therefore if we fix Y and Z the right hand side of (4) is a 1-form ω and we may define $\nabla_Z Y$ to be $\frac{1}{2}\, g^{-1}(\omega)$. Then $\nabla_Z Y$ is F-linear in Z, for if we replace Z by fZ then f is differentiated via

$$(Y\cdot f)g(Z,X) - (X\cdot f)g(Y,Z) + (X\cdot f)g(Z,Y) - (Y\cdot f)g(Z,X) = 0 .$$

Finally, if we replace Y by fY then f is differentiated via

$$(Z\cdot f)g(X,Y) - (X\cdot f)g(Y,Z) + (Z\cdot f)g(Y,X) + (X\cdot f)g(Y,Z) = 2(Z\cdot f)g(X,Y)$$

so that

$$\nabla_Z(fY) = f\nabla_Z Y + (Z\cdot f)Y .$$

Thus ∇ is an affine connection.

Let $e(X,Y,Z)$ be the left hand side of (1) minus the right hand side of (1). We have seen that (4) is equivalent to

$$e(X,Y,Z) - e(Y,Z,X) + e(Z,X,Y) = 0 . \tag{5}$$

Therefore (5) is true, and by cyclic permutation

$$e(Y,Z,X) - e(Z,X,Y) + e(X,Y,Z) = 0 . \tag{6}$$

By (5) and (6), $e(X,Y,Z) = 0$. That is, (1) holds.

Let T be the torsion of ∇. By (1),

$$Z\cdot g(X,Y) = g(\nabla_Z X,Y) + g(T(X,Z),Y) + g(X,\nabla_Z Y) . \tag{7}$$

We also have

$$Z\cdot g(X,Y) = (\nabla_Z g)(X,Y) + g(\nabla_Z X,Y) + g(X,\nabla_Z Y) . \tag{8}$$

By (7) and (8)

$$(\nabla_Z g)(X,Y) = g(T(X,Z),Y) . \tag{9}$$

We need to show that both sides of (9) are 0. Let $u(X,Y,Z)$ denote the common value. The u is symmetric in the first two variables (by the left hand side of (9)) and antisymmetric in the first and third variables (by the right hand side of (9)). Therefore

$$\begin{aligned} u(X,Y,Z) &= -u(Z,Y,X) = -u(Y,Z,X) = u(X,Z,Y) \\ &= u(Z,X,Y) = -u(Y,X,Z) = -u(X,Y,Z) \end{aligned}$$

and $u = 0$. Thus $\nabla g = 0$ and since g is non-degenerate, $T = 0$. QED

The connection ∇ of Theorem 1 is called the Riemannian connection or the Levi-Civita connection. It is due to Christoffel.

Theorem 2. *Let g be a pseudo-Riemannian metric on a coordinate Lie module E with coordinates $x^1,\dots,x^n$. Then the components of the Riemannian connection ∇ are given by*

$$\Gamma^k_{ij} = \frac{1}{2}\Sigma g^{ka}\left(\frac{\partial}{\partial x^i} g_{aj} + \frac{\partial}{\partial x^j} g_{ai} - \frac{\partial}{\partial x^a} g_{ij}\right) . \tag{10}$$

Proof. Apply (4) to $X = \partial/\partial x^a$, $Z = \partial/\partial x^i$, $Y = \partial/\partial x^j$. Then

$$2\Sigma g_{ba}\Gamma^b_{ij} = \frac{\partial}{\partial x^i} g_{aj} + \frac{\partial}{\partial x^j} g_{ai} - \frac{\partial}{\partial x^a} g_{ij} .$$

If we multiply by g^{ka} and sum over a we obtain (10). QED.

If E is merely free of finite type we obtain three additional terms involving the structure scalars due to the fact that the vector fields of the basis may not commute.

The discovery of the Christoffel symbols (10) and of covariant differentiation was a marvellous discovery. Despite the fact that tensor analysis has been studied so intensively that it has become easy it remains an amazing subject.

3. Raising and lowering indices

Let g be a pseudo-Riemannian metric on the F module E and let u be a covariant tensor of rank r. Let $1\leq\mu\leq r$. Then we define a tensor v, contravariant of rank 1 and covariant of rank $r-1$ by

$$v(\omega,X_1,\ldots,\hat{X}_\mu,\ldots,X_r) = u(X_1,\ldots,X_{\mu-1},g^{-1}(\omega),X_{\mu+1},\ldots,X_r) .$$

In the classical tensor notation (§1.12) this is indicated by

$$u_{i_1\ldots i_{\mu-1}i_{\mu+1}\ldots i_r}^{\phantom{i_1\ldots i_{\mu-1}}i_\mu}$$

(where $i_\mu = \omega$ and $i_\nu = X_\nu$ for $\nu \neq \mu$). This is a notation of unsurpassable clarity. Notice that it is important to leave a space below the raised index to indicate precisely which tensor in E_1^{r-1} correesponding to the given tensor in E^r is meant. Notice that we may extend the notation to allow several raised indices. It is customary when working with a fixed pseudo-Riemannian metric g to write all tensors with the upper and lower indices ordered relative to each other to indicate the relation among the various tensors formed by raising and lowering indices. If E is totally reflexive, so that contractions of tensors make sense, we continue to indicate contractions by the use of dummy indices as before.

4. The Riemann-Christoffel tensor

Let g be a pseudo-Riemannian metric on the Lie module E, ∇ the Riemannian connection and R its curvature. We define a tensor $\tilde{R}$, covariant of rank 4 called the Riemann-Christoffel tensor by

$$\tilde{R}(W,Z,X,Y) = g(W,R(X,Y)Z) .$$

Thus $\tilde{R}(W,Z,X,Y) = R(g(W),Z,X,Y)$ where R is the curvature tensor. In the classical tensor notation we denote the values of the curvature tensor by

$$R^i{}_{jk\ell} \ .$$

Thus the Riemann-Christoffel tensor is obtained by lowering the contravariant index i, and its values are denoted

$$R_{ijk\ell} \ .$$

(Notice that no tilde is necessary.) For the geometric meaning of the Riemann-Christoffel tensor see pp. 64-70 of Helgason [2,§2].

<u>Theorem 3.</u> <u>Let</u> g <u>be a pseudo-Riemannian metric on the Lie module</u> E . <u>Then the Riemann-Christoffel tensor satisfies</u>

(11) $$R_{ij\ell k} = -R_{ijk\ell} \ ,$$

(12) $$R_{ijk\ell} + R_{ik\ell j} + R_{i\ell jk} = 0 \ ,$$

(13) $$R_{jik\ell} = -R_{ijk\ell} \ ,$$

(14) $$R_{k\ell ij} = R_{ijk\ell} \ .$$

<u>Proof.</u> We have already seen (§5.10) that if R is the curvature tensor of any affine connection then

$$R^i{}_{j\ell k} = -R^i{}_{jk\ell} \ ,$$

$$R^i{}_{j\ell k} + R^i{}_{\ell kj} + R^i{}_{kj\ell} = 0 \ .$$

Therefore (11) and (12) hold. The equation (13) may be re-written as

(15) $$g(W,R(X,Y)Z) = -g(R(X,Y)W,Z) \ .$$

(The equation (15) says that each $R(X,Y)$ is a skew-symmetric operator with respect to g, which means, formally, that it is the derivative at $t=0$ of a one-parameter group of g-isometries.) To prove (15), we apply $[\nabla_X,\nabla_Y] - \nabla_{[X,Y]}$ to $g(W,Z)$, obtaining

$$0 = ((\nabla_X\nabla_Y - \nabla_Y\nabla_X - \nabla_{[X,Y]})g)(W,Z) + g(R(X,Y)W,Z) + g(W,R(X,Y)Z)$$

by §5.11. (The left hand side is 0 since $g(W,Z)$ is a scalar.) The first term on the right hand side is 0 since $\nabla g = 0$. Therefore (13) holds.

The equation (14) is a consequence of (11), (12), and (13), as follows. By (12),

$$\begin{aligned} &R_{ijk\ell} + R_{ik\ell j} + R_{i\ell jk} \\ &+ R_{jik\ell} + R_{jk\ell i} + R_{j\ell ik} \\ &+ R_{kji\ell} + R_{ki\ell j} + R_{k\ell ji} \\ &+ R_{\ell jki} + R_{\ell kij} + R_{\ell ijk} = 0 \end{aligned}$$

since each line is 0. By (13) the top row cancels with the main diagonal of the remaining nine terms. By (11) and (13) the remaining six terms are symmetric about the main diagonal, so that

(16) $$R_{jk\ell i} + R_{j\ell ik} + R_{k\ell ji} = 0 .$$

By (12) the first two terms of (16) add up to $-R_{jik\ell}$ so that

$$-R_{jik\ell} + R_{k\ell ji} = 0 .$$

Thus (14) holds. QED.

Let us suppose that E is totally reflexive, so that tensors may be contracted. By (13), $R^i{}_{ik\ell} = 0$. This property is not shared by all affine connections: it says that the trace of each $R(X,Y)$ is 0. Consider however

$$R_{jk} = R^{i}{}_{jki} .$$

This is a symmetric (by (11), (13), and (14)) covariant tensor of rank 2, called the Ricci curvature. The corresponding tensor

$$R_j{}^k = R^{i}{}_{j}{}^{k}{}_{i} \tag{17}$$

in E^1_1 is called the Ricci tensor. Since the Ricci curvature is symmetric, the Ricci tensor may be written as R^k_j without danger of confusion. Its trace R^j_j is called the scalar curvature. For those who wish to avoid indices the Ricci tensor (17) may be written as

$$C^1_2(g^{-1} \otimes C^1_3 R) ,$$

a notation of almost magical inefficiency.

5. The codifferential

Let ∇ be the Riemannian connection with respect to a pseudo-Riemannian metric g on a totally reflexive Lie module E. If α is an r-form we define

$$(\delta\alpha)_{i_1 \dots i_{r-1}} = -\nabla^a \alpha_{a i_1 \dots i_{r-1}} , \tag{18}$$

and $\delta\alpha = 0$ if α is a 0-form. Recall the convention (12) of §1.12, that for α in A^r,

$$\alpha_{i_1 \dots i_r} = r! \alpha(i_1, \dots, i_r) .$$

The (r-1)-form $\delta\alpha$ is called the codifferential of α, and α is called co-closed if $\delta\alpha = 0$, co-exact if for some β we have $\delta\beta = \alpha$.

The operator δ is called the codifferential. It was introduced in 1923 by Weitzenböck [6], to whom the following theorem and proof are due.

Theorem 4. *Let δ be the codifferential, with respect to a pseudo-Riemannian metric g on a totally reflexive Lie module E. Then $\delta^2 = 0$.*

Proof. Let $\alpha \in A^r$. Then

$$(\delta\delta\alpha)_{i_1\ldots i_{r-2}} = -\nabla^a(\delta\alpha)_{ai_1\ldots i_{r-2}} = \nabla^a\nabla^b\alpha_{bai_1\ldots i_{r-2}} .$$

By Ricci's identity (§5.11)

$$(19)\qquad \nabla_j\nabla_k\alpha_{bai_1\ldots i_{r-2}} = \nabla_k\nabla_j\alpha_{bai_1\ldots i_{r-2}} + R^c{}_{bkj}\alpha_{cai_1\ldots i_{r-2}} + R^c{}_{akj}\alpha_{bci_1\ldots i_{r-2}} + \sum_{\nu=1}^{r-2} R^c{}_{i_\nu kj}\alpha_{bai_1\ldots i_{\nu-1}ci_{\nu+1}\ldots i_{r-2}} .$$

Now raise the indices j and k and contract with a and b respectively. The left hand side of (19) gives $\delta^2\alpha$, the next term gives $-\delta^2\alpha$ due to the antisymmetry of α in b and a. Since $R^c{}_b{}^{ba}$ and $R^c{}_a{}^{ba}$ are symmetric tensors, the next two terms give 0 for the same reason. Therefore

$$(20)\qquad 2(\delta^2\alpha)_{i_1\ldots i_{r-2}} = \sum_{\nu=1}^{r-2} R^c{}_{i_\nu}{}^{ba}\alpha_{bai_1\ldots i_{\nu-1}ci_{\nu+1}\ldots i_{r-2}} .$$

Cyclic permutations of c, b, and a leave α unchanged, but

$$R^c{}_i{}^{ba} + R^b{}_i{}^{ac} + R^a{}_i{}^{cb} = 0$$

by (13) and (12), so that (20) is 0. QED.

Modern treatments (see deRham [7]) give a much shorter proof of this theorem by use of the operator $*$. However, the introduction of $*$ leads to needless complications related to orientability. Furthermore, the above definition of δ and proof that $\delta^2 = 0$ generalize to infinite dimensional pseudo-Riemannian manifolds (and could conceivably be of interest), provided one pays attention to convergence problems when taking contractions. There is no $*$ on an infinite dimensional manifold.

6. Divergences

Let g be a pseudo-Riemannian metric on a totally reflexive F module E. We define $g(u,v)$ for u and v in E^s_r by

$$g(u,v) = u^{j_1 \cdots j_s}_{i_1 \cdots i_r} v^{i_1 \cdots i_r}_{j_1 \cdots j_s}$$

(where we have not indicated relative order of contravariant and covariant indices). As with an r-form α itself, we make the <u>convention</u> that α with raised indices is $r!$ times what it would be if regarded simply as a tensor, and we make the <u>convention</u> that for α and β in A^r,

$$g(\alpha,\beta) = \frac{1}{r!} \alpha^{i_1 \cdots i_r} \beta_{i_1 \cdots i_r} .$$

If $u \in E^s_r$ and $v \in E^{s'}_{r'}$ where $(r',s') \neq (r,s)$ we define $g(u,v) = 0$, and we extend $g(u,v)$ by additivity in each variable to all of the mixed tensor algebra. Similarly we define $g(\alpha,\beta)$ for arbitrary exterior forms α and β.

If ∇ is the Riemannian connection with respect to a pseudo-Riemannian metric on a totally reflexive Lie module and ω is a 1-form, we call $\nabla^i \omega_i = -\delta\omega$ the divergence of ω, and any scalar of this form is a divergence.

Theorem 5. Let δ be the codifferential, with respect to a pseudo-Riemannian metric g on a totally reflexive Lie module E. Then for all exterior forms α and β

$$g(d\alpha, \beta) - g(\alpha, \delta\beta)$$

is a divergence.

Proof. It suffices to prove this for α in A^r and β in A^{r+1}. Let us write $f \equiv g$ in case $f-g$ is a divergence. Recall that since the Riemannian connection is torsion-free, we have (§5.7)

$$(d\alpha)_{i_1 \dots i_{r+1}} = \sum_{\mu=1}^{r+1} (-1)^{\mu+1} \nabla_{i_\mu} \alpha_{i_1 \dots \hat{i}_\mu \dots i_{r+1}} . \tag{21}$$

Therefore

$$g(d\alpha, \beta) = \frac{1}{(r+1)!} \sum_{\mu=1}^{r+1} (-1)^{\mu+1} (\nabla^{i_\mu} \alpha^{i_1 \dots \hat{i}_\mu \dots i_{r+1}}) \beta_{i_1 \dots i_{r+1}}$$

$$= \frac{1}{(r+1)!} \sum_{\mu=1}^{r+1} (-1)^{\mu+1} \{-\alpha^{i_1 \dots \hat{i}_\mu \dots i_{r+1}} \nabla^{i_\mu} \beta_{i_1 \dots i_{r+1}}$$

$$+ \nabla^{i_\mu} (\alpha^{i_1 \dots \hat{i}_\mu \dots i_{r+1}} \beta_{i_1 \dots i_{r+1}})\}$$

$$\equiv \frac{1}{(r+1)!} \sum_{\mu=1}^{r+1} (-1)^{\mu+1} \{-\alpha^{i_1 \dots \hat{i}_\mu \dots i_{r+1}} \nabla^{i_\mu} \beta_{i_1 \dots i_{r+1}}\}$$

$$= \frac{1}{r!} \alpha^{i_1 \dots i_r} (\delta\beta)_{i_1 \dots i_r} = g(\alpha, \delta\beta) .$$

QED.

This theorem says that δ is the adjoint operator to d. Notice that quite apart from any question of notational conventions, it is necessary to define the g-inner product on forms by weighting forms of various degrees differently in order for this to be the case.

7. The Laplace operator

If f is a scalar so is $\nabla^i\nabla_i f$. The operator $\nabla^i\nabla_i$ on scalars is called the Laplace-Beltrami operator, and the same terminology may be used for the operator $\nabla^i\nabla_i$ on arbitrary tensors. Notice that for f a scalar,

$$\nabla^i\nabla_i f = -\delta df = -(\delta d + d\delta)f .$$

It is natural to study the operator $-(\delta d + d\delta)$ on exterior forms, and this was first done by Weitzenböck. Unfortunately, the wrong sign convention has been adopted in all recent accounts of harmonic integrals and one refers to $\delta d + d\delta$ as the Laplacean!

The Laplace operator in its various manifestations is the most beautiful and central object in all of mathematics. Probability theory, mathematical physics, Fourier analysis, partial differential equations, the theory of Lie groups, and differential geometry all revolve around this sun, and its light even penetrates such obscure regions as number theory and algebraic geometry. Only with pain do I adopt the sign convention which is standard in the theory of harmonic forms.

Definition. Let g be a pseudo-Riemannian metric on a totally reflexive Lie module. The Laplace-deRham operator Δ on A^* is defined by

$$\Delta = \delta d + d\delta .$$

An exterior form α is harmonic in case $\Delta\alpha = 0$.

Notice that Δ maps each A^r into itself and that α is harmonic if and only if each of its homogeneous components is. The set $\mathcal{H}$ of harmonic forms is a graded F^0 vector space, and its homogeneous subspaces are denoted $\mathcal{H}^r$.

8. The Weitzenböck formula

Theorem 6. Let Δ be the Laplace-deRham operator, with respect to a pseudo-Riemannian metric on a totally reflexive Lie module. For all r-forms α,

(22) $$(\Delta\alpha)_{i_1\ldots i_r} = -\nabla^a\nabla_a\alpha_{i_1\ldots i_r} + \sum_{\nu=1}^{r}(-1)^\nu R^b{}_a{}^a{}_{i_\nu}\alpha_{bi_1\ldots\hat{i}_\nu\ldots i_r} + \sum_{\nu<\mu}(-1)^{\mu+\nu}R^b{}_{i_\nu}{}^a{}_{i_\mu}\alpha_{abi_1\ldots\hat{i}_\nu\ldots\hat{i}_\mu\ldots i_r}.$$

Proof. By the formulas (18) and (21) for d and δ,

$$(d\delta\alpha)_{i_1\ldots i_r} = \sum_{\nu=1}^{r}(-1)^\nu\nabla_{i_\nu}\nabla^a\alpha_{ai_1\ldots\hat{i}_\nu\ldots i_r}$$

and

$$(\delta d\alpha)_{i_1\ldots i_r} = -\nabla^a(d\alpha)_{ai_1\ldots i_r} = -\nabla^a\nabla_a\alpha_{i_1\ldots i_r} - \sum_{\nu=1}^{r}\nabla^a\nabla_{i_\nu}\alpha_{ai_1\ldots\hat{i}_\nu\ldots i_r},$$

so that

$$(\Delta\alpha)_{i_1\ldots i_r} = -\nabla^a\nabla_a\alpha_{i_1\ldots i_r} + \sum_{\nu=1}^{r}(-1)^\nu(\nabla_{i_\nu}\nabla^a - \nabla^a\nabla_{i_\nu})\alpha_{i_1\ldots\hat{i}_\nu\ldots i_r}.$$

If we use Ricci's identity (§5.11) and the fact that α is alternating, we see that this is equal to

$$-\nabla^a\nabla_a\alpha_{i_1\ldots i_r} + \sum_{\nu=1}^{r}(-1)^\nu\{R^b{}_a{}^a{}_{i_\nu}\alpha_{bi_1\ldots\hat{i}_\nu\ldots i_r} + R^b{}_{i_1}{}^a{}_{i_\nu}\alpha_{abi_2\ldots\hat{i}_\nu\ldots i_r} + \ldots + R^b{}_{i_r}{}^a{}_{i_\nu}\alpha_{ai_1\ldots\hat{i}_\nu\ldots i_{r-1}b}\}$$

$$= -\nabla^a\nabla_a\alpha_{i_1\ldots i_r} + \sum_{\nu=1}^{r}(-1)^\nu R^b{}_a{}^a{}_{i_\nu}\alpha_{bi_1\ldots\hat{i}_\nu\ldots i_r} + 2\sum_{\nu<\mu}(-1)^{\mu+\nu}R^b{}_{i_\mu}{}^a{}_{i_\nu}\alpha_{abi_1\ldots\hat{i}_\nu\ldots\hat{i}_\mu\ldots i_r}.$$

QED.

The classical tensor notation is a convenient and flexible computational device, but it is not immediately clear what the meaning of the additional terms in the Weitzenböck formula is. To reformulate the formula, let $\bar{R}$ be the tensor in E^2_2 given by raising the second covariant index in the curvature tensor R (so that $\bar{R}$ has values $R^{i\ k}_{\ j\ \ell}$; that is, $\bar{R}(\omega,\eta,X,Y) = \langle \omega, R(g^{-1}\eta,Y)X \rangle$) and recall the operator $\Phi_{\bar{R}}$ of §3.4.

<u>Theorem 7. Let</u> Δ <u>be the Laplace-deRham operator, with respect to a pseudo-Riemannian metric on a totally reflexive Lie module. For all exterior forms</u> α ,

$$\Delta\alpha = -\nabla^a\nabla_a\alpha + \Phi_{\bar{R}}\alpha \ .$$

<u>Proof.</u> Compare (11) of §3.4 with (22) and recall that

$$R^{b\ \ a}_{\ i_\nu\ b} = R^{a\ b}_{\ b\ i_\nu}$$

by (14). QED.

9. Operators commuting with the Laplacean

From the definition $\Delta = d\delta + \delta d$ of the Laplace-deRham operator and the fact that $d^2 = \delta^2 = 0$, it is clear that d and δ commute with Δ . Here we shall investigate F-linear mappings of A^* into A^* which commute with Δ .

Let E be a totally reflexive F module, L an F-linear mapping of A^* into A^* . Let $L_{(r,s)}\alpha$ be the component in A^r of $L\beta$, where β is the component in A^s of α . Then $L_{(r,s)}$ is also an F-linear mapping of A^* into A^* . Now let $L = L_{(r,s)}$. By §1.6, there is a tensor L in E^r_s such that

$$(L\alpha)_{i_1\cdots i_r} = L_{i_1\cdots i_r}^{j_1\cdots j_s}\alpha_{j_1\cdots j_s}, \qquad \alpha \in A^s .$$

Since $L\alpha \in A^r$ for all α in A^s, the tensor L is alternating in its covariant indices, and there is no loss of generality in assuming it to be alternating in its contravariant indices. Thus the set of F-linear mappings of A^s into A^r may be identified with

$$A_s^r = A^r \otimes A_s$$

where A_s is the elements of rank s in the contravariant Grassman algebra A_*. Let A_*^* be the strong direct sum of the A_s^r. (Of course, if $A^r = 0$ for r sufficiently large, this is the same as the weak direct sum.) Then we may identify F-linear mappings of A^* into itself with the elements of A_*^*. If E is a Lie module, ∇ an affine connection, and L an F-linear mapping of A^* into itself we define $\nabla L = 0$ to mean that for each tensor $L_{(r,s)}$ in A_s^r we have $\nabla L_{(r,s)} = 0$.

Recall the definition at the end of §1 of the notion of a punctual F module.

Theorem 8. Let Δ be the Laplace-deRham operator, with respect to a pseudo-Riemannian metric on a totally reflexive punctual Lie module E. Let L be an F-linear mapping of the exterior forms A^* into A^* with covariant derivative 0 (with respect to the Riemannian connection). Then L commutes with Δ.

Proof. It suffices to prove this for $L = L_{(r,s)}$; that is, $L\colon A^s \longrightarrow A^r$ and $L\colon A^{s'} \longrightarrow 0$ for $s' \neq s$. By Theorem 4 of §3.2, $[\nabla_X, L] = 0$ for all vector fields X. Therefore

$$(\nabla^a\nabla_a L\alpha)_{i_1\cdots i_r} = \nabla^a\nabla_a L_{i_1\cdots i_r}^{j_1\cdots j_s}\alpha_{j_1\cdots j_s}$$

$$= L_{i_1\cdots i_r}^{j_1\cdots j_s}\nabla^a\nabla_a\alpha_{j_1\cdots j_s} = (L\nabla^a\nabla_a\alpha)_{i_1\cdots i_r}, \qquad \alpha \in A^s.$$

If $\alpha \in A^{s'}$ for $s' \neq s$ then $\nabla^a\nabla_a\alpha$ is also in $A^{s'}$, so that $\nabla^a\nabla_a L\alpha$ and $L\nabla^a\nabla_a\alpha$ are both 0. Consequently, L commutes with the Laplace-Beltrami operator $\nabla^a\nabla_a$. By Theorem 7, we need only show that L commutes with $\Phi_{\overline{R}}$.

Since the Riemannian connection ∇ is torsion-free, we have by §5.11 that

$$\varphi_{R(X,Y)} = \nabla_X\nabla_Y - \nabla_Y\nabla_X - \nabla_{[X,Y]},$$

so that by Theorem 4 of §3.2 again, L commutes with each $\varphi_{R(X,Y)}$.

Suppose there is an s-form α such that

$$\beta = L\Phi_{\overline{R}}\alpha - \Phi_{\overline{R}}L\alpha$$

is not 0. Then there is a γ in E_r (in fact, we may take γ in A_r) such that the scalar $<\beta,\gamma>$ is not 0, since E is totally reflexive. Since E is punctual, there is a homomorphism

$$\rho\colon (F,E,E') \longrightarrow (F^0,W,W'),$$

where W is a finite-dimensional vector space over F^0, such that $\rho<\beta,\gamma> \neq 0$. The homomorphism ρ induces a homomorphism ρ of the mixed tensor algebra E_*^* into the mixed tensor algebra W_*^*, and $\rho<\beta,\gamma> = <\rho\beta,\rho\gamma>$. Therefore, $\rho\beta \neq 0$. Consequently we need only prove that

$$\rho(L)\rho(\Phi_{\overline{R}}) = \rho(\Phi_{\overline{R}})\rho(L).$$

Let V be the vector subspace of W_1^1 spanned by all $\rho(R(X,Y))$. Since L commutes with each $\varphi_{R(X,Y)}$, $\rho(L)$ commutes with each $\rho(\varphi_{R(X,Y)})$ and so $\rho(L)$ commutes with φ_A for each A in V.

For each η in E' and Y in E, we have $\overline{R}(\cdot,\eta,\cdot,Y) = R(g^{-1}\eta,Y)$, so that $\rho(\overline{R}(\cdot,\eta,\cdot,Y) = \rho(R(g^{-1}\eta,Y))$ is in V. By the symmetry (14), each $\rho(\overline{R}(\omega,\cdot,X,\cdot))$ is in V. Now $\rho(\overline{R})$ is in W_2^2 and $W_2^2 = W_1^1 \otimes W_1^1$. Let $e_1,\ldots,e_n$ be a basis of the finite-dimensional vector W_1^1 such that the first m of them $(m \leq n)$ are a basis of V, and let $f^1,\ldots,f^n$ be the dual basis. Let

$$\rho(\overline{R}) = \Sigma c^{ij} e_i \otimes e_j , \qquad c^{ij} \in F^o .$$

Then we have $c^{ij} = \langle f^i \otimes f^j, \rho(\overline{R}) \rangle$, and by what we have just seen this is 0 if $i > m$ or $j > m$. By the definition of Φ (§3.4) we have

$$\rho(\Phi_{\overline{R}}) = \Phi_{\rho(\overline{R})} = \sum_{i,j=1}^{m} c^{ij} \varphi_{e_i} \varphi_{e_j} .$$

Since e_i, e_j are in V for $i,j \leq m$ this means that $\rho(L)$ commutes with $\rho(\Phi_{\overline{R}})$. QED.

Notice that a mapping L with $\nabla L = 0$ need not commute with d or δ. For example, let L be the mapping which sends each exterior form α into its component in A^r. Then $\nabla L = 0$ for any affine connection ∇ but dL is d on A^r and Ld is 0 on A^r.

If g is a pseudo-Riemannian metric on a totally reflexive punctual Lie module E, we denote by $\mathcal{Q}$ the set of all F-linear mappings L of A^* into itself which have covariant derivative 0. Then $\mathcal{Q}$ is clearly an F^o algebra. The algebra $\mathcal{Q}$ possesses a natural involution $*$ defined as follows. If $L \in A_s^r$ then L^* in A_r^s is given by

$$(L^*)^{i_1\cdots i_r}_{j_1\cdots j_s} = L^{j_1\cdots j_s}_{i_1\cdots i_r} ,$$

and $*$ is extended to A^*_* by additivity. Then $*$ maps $\mathcal{Q}$ into itself, and

$$(L+M)^* = L^* + M^* ,$$
$$L^{**} = L ,$$
$$(LM)^* = M^* L^* .$$

By Theorem 8 each L in $\mathcal{Q}$ maps the harmonic forms into themselves, so that by restriction we obtain a $*$-representation of the bi-graded algebra $\mathcal{Q}$ with involution on the graded vector space $\mathcal{H}^*$.

10. Hodge theory

If M is a differentiable manifold (paracompact and finite dimensional) then we may construct a Riemannian metric g on it. This is trivial to do locally, and then we construct g globally using a partition of unity. Since a convex combination of Riemannian metrics is again a Riemannian metric, there is no difficulty.

A manifold M, orientable or not, with a pseudo-Riemannian metric g has a distinguished volume element (a measure, not an n-form) which we denote dV. In local coordinates,

$$dV = \sqrt{|\det g_{ij}|}\, dx^1 \ldots dx^n .$$

If M is compact then the integral of any scalar which is a divergence is 0.

For the rest of this discussion, let M be a compact Riemannian manifold. Then the exterior forms A^* form a pre-Hilbert space if we define the (real-valued) inner product of two forms α and β to be

$$(\alpha,\beta) = \int_M g(\alpha,\beta)dV \ .$$

By Theorem 5,

$$(d\alpha,\beta) = (\alpha,\delta\beta) \ , \qquad \alpha,\beta \in A^* \ .$$

Therefore, if α is harmonic

$$\begin{aligned} 0 = (\Delta\alpha,\alpha) &= (d\alpha,d\alpha) + (\delta\alpha,\delta\alpha) \\ &= \int_M g(d\alpha,d\alpha)dV + \int_M g(\delta\alpha,\delta\alpha)dV \ . \end{aligned}$$

Since the integrands are positive and continuous they are 0. Consequently, on a compact Riemannian manifold a harmonic form is closed and co-closed. (The converse is trivially true on any pseudo-Riemannian manifold.) This is why the Laplace-deRham operator Δ was introduced. (It was introduced in Hodge's theory of harmonic integrals by Kodaira and independently by Bidal and deRham.)

The operator Δ is a symmetric, positive operator on the pre-Hilbert space A^*. As a partial differential operator it is elliptic, for by the Weitzenböck formula in local coordinates Δ is $-g^{ij}\partial^2/\partial x^i\partial x^j$ plus lower order terms, and the matrix g^{ij} is of strictly positive type. It follows that the closure of the operator Δ (which we again denote Δ) is a self-adjoint, positive operator on the completion of the pre-Hilbert space A^*, and Δ has discrete spectrum.

Let

$$h(t) = \begin{cases} 1 \ , & t = 0 \\ 0 \ , & t \neq 0 \ , \end{cases}$$

$$g(t) = \begin{cases} 0 \ , & t = 0 \\ \frac{1}{t} \ , & t \neq 0 \ , \end{cases}$$

and define $H = h(\Delta)$, $G = g(\Delta)$ by the functional calculus for self-adjoint operators on Hilbert space. They are bounded self-adjoint operators, and H is the orthogonal projection onto the null-space of Δ. By the regularity theorem for elliptic operators, H and G map A^* into itself. In fact, H maps the entire Hilbert space into A^*, so that $H\alpha = \alpha$ if and only if α is harmonic.

Since d and δ commute with Δ, and H and G are functions of Δ, d and δ commute with H and G. From the definition of G, $\Delta G = d\delta G + \delta d G$ is the orthogonal projection $1-H$ onto the orthogonal complement of the harmonic forms. Therefore we have the decomposition for any α in A^*,

$$\alpha = d\delta G\alpha + \delta d G\alpha + H\alpha .$$

If α is closed, the second term $\delta dG\alpha = \delta G d\alpha$ is 0, so that a closed form is cohomologous to its harmonic part $H\alpha$. If α is exact, $\alpha = d\beta$, then its harmonic part is 0, since $H\alpha = Hd\beta = dH\beta = 0$. Therefore we have the Hodge theorem: On a compact Riemannian manifold every harmonic form is closed and co-closed, and the mapping which sends a harmonic form into its cohomology class is a vector space isomorphism of $\mathcal{H}^*$ onto H^*.

The Hodge theorem is of great importance in Riemannian geometry. It can sometimes be used to compute cohomology groups but its chief importance lies in the possibility it affords of deducing global restrictions on the topology of a manifold in order that it admit certain types of differential-geometric structures.

By Theorem 8 the algebra $\mathcal{R}$ of F-linear mappings of A^* into itself with covariant derivative 0 acts on $\mathcal{H}^*$ and therefore, by the Hodge theorem, on H^*. This places restrictions on the topology of a compact n-manifold which admits a Riemannian metric with a given subgroup of $O(n, \mathbb{R})$ as holonomy group.

Theorem 8 is due to Weil, and a number of examples are well-known. Suppose M is an orientable compact Riemannian n-manifold. Then there is an F-linear operator $*$ mapping A^r isomorphically onto A^{n-r} and commuting with Δ. By the Hodge theorem, it induces an isomorphism of H^r onto H^{n-r}. This is a weak form of Poincaré duality (weak because we have cohomology with real coefficients). The principal application of Hodge theory is to the topology of Kähler manifolds. If Ω is the symplectic 2-form with covariant derivative 0 then $L\alpha = \Omega \wedge \alpha$ and $\Lambda = L^*$ commute with Δ and so induce operators on H^*. The existence of these operators places strong restrictions on the cohomology of a Kähler manifold.

The application of Hodge theory requires the geometrical structure under study to be related to a Riemannian metric. One of the main open fields of research in differential geometry is the problem of finding global restrictions on a manifold in order for it to admit a more general geometrical structure, such as a foliation or complex structure, satisfying integrability conditions.

References

[5a] A. Weil, Un théorème fondamental de Chern en géometrie riemannienne, Seminaire Bourbaki, No. 239, 1961-1962.

[6] R. Weitzenböck, Invariantentheorie, Noordhoff, Groningen, 1923.

[7] Georges deRham, Variétés differentiables - Formes, Courants, Formes harmonique, Hermann, Paris, 1960.

§8. Symplectic structures

1. Almost symplectic structures

A 2-form Ω is in particular a covariant tensor of rank 2 and so determines a mapping $\Omega\colon E \longrightarrow E'$ given by $\langle \Omega(X), Y\rangle = \Omega(X,Y)$, and is called non-degenerate if the mapping is bijective (§7.1).

Definition. An almost symplectic structure on an F module E is a non-degenerate 2-form.

Thus the definition is the same as that of a pseudo-Riemannian metric except for a minus sign. If E is free of finite type with basis $X_1,\ldots,X_n$ and Ω is an almost symplectic structure with components Ω_{ij} then clearly $\Omega_{ji} = -\Omega_{ij}$ and $\det \Omega_{ij} \neq 0$, and conversely. Also, n must be even since

$$\det \Omega_{ij} = \det \Omega_{ji} = \det(-1) \det \Omega_{ij} = (-1)^n \det \Omega_{ij} .$$

Definition. A symplectic structure on a Lie module E is a closed almost symplectic structure.

That is, a non-degenerate 2-form Ω which satisfies $d\Omega = 0$ is called a symplectic structure. As in §7, we begin by studying affine connections which preserve the structure.

Theorem 1. Let Ω be an almost symplectic structure on the Lie module E. If ∇ is an affine connection such that $\nabla\Omega = 0$ then

$$3d\Omega(X,Y,Z) = \mathfrak{S}\, \Omega(T(X,Y),Z) .$$

In particular, if there is a torsion-free affine connection ∇ such that $\nabla\Omega = 0$ then Ω is a symplectic structure.

Proof. Since $\nabla\Omega = 0$ we have

$$X\cdot\Omega(Y,Z) = \Omega(\nabla_X Y,Z) + \Omega(Y,\nabla_X Z)$$
$$= \Omega(\nabla_Y X,Z) + \Omega([X,Y],Z) + \Omega(T(X,Y),Z) + \Omega(Y,\nabla_X Z)$$

by the definition of torsion (§5.6). When we take cyclic sums the first and last terms of the right hand side cancel, so that

$$\mathfrak{S}\{X\cdot\Omega(Y,Z) - \Omega([X,Y],Z)\} = \mathfrak{S}\,\Omega(T(X,Y),Z) .$$

As we remarked before (formula (8), §4.2), the left hand side of this is $3d\Omega(X,Y,Z)$. The last statement in the theorem follows from this or the remark that $d\Omega = \mathrm{Alt}\,\nabla\Omega$ for ∇ torsion-free. QED.

2. Hamiltonian vector fields and Poisson brackets

Let Ω be a symplectic structure on the Lie module E. If h is a scalar then $\Omega^{-1}dh$ is a vector field, and a vector field of this form is called Hamiltonian. That is, X is a Hamiltonian vector field in case ΩX is exact. A vector field X such that ΩX is closed is called locally Hamiltonian (because by Poincaré's lemma a closed form on a manifold is locally exact). We may transport the Lie product to the 1-forms as follows: if ω_1 and ω_2 are 1-forms their Poisson bracket is

$$[\omega_1,\omega_2] = \Omega[\Omega^{-1}\omega_1,\Omega^{-1}\omega_2] .$$

If f and g are scalars we define their Poisson bracket to be

$$[f,g] = -\Omega(\Omega^{-1}\,df,\Omega^{-1}\,dg) .$$

Theorem 2. Let Ω be a symplectic structure on the Lie module E. Then:

(a) <u>A vector field</u> X <u>is locally Hamiltonian if and only if</u> $\theta_X \Omega = 0$.

(b) <u>If</u> ω_1 <u>is closed then</u> $[\omega_1,\omega_2] = \theta_{\Omega^{-1}\omega_1} \omega_2$.

(c) $[f,g] = (\Omega^{-1}\, df)\cdot g = -(\Omega^{-1}\, dg)\cdot f$.

(d) $[df,dg] = d[f,g]$.

<u>Proof.</u> By Theorem 4 of §4.3, $\theta_X \Omega = d(X \lrcorner \Omega) + X \lrcorner d\Omega$. Since Ω is closed, $\theta_X \Omega = \frac{1}{2}\, d\Omega X$, which proves (a). If ω_1 is closed then $\Omega^{-1}\omega_1$ is locally Hamiltonian, so by (a)

$$\theta_{\Omega^{-1}\omega_1} \omega_2 = \theta_{\Omega^{-1}\omega_1} \Omega\Omega^{-1}\omega_2$$

$$= \Omega\theta_{\Omega^{-1}\omega_1} \Omega^{-1}\omega_2 = \Omega[\Omega^{-1}\omega_1, \Omega^{-1}\omega_2] = [\omega_1,\omega_2],$$

which proves (b). To prove (c), notice that

$$[f,g] = -\Omega(\Omega^{-1}\, df, \Omega^{-1}\, dg) = -\langle \Omega\Omega^{-1}\, df, \Omega^{-1}\, dg \rangle$$

$$= -\langle df, \Omega^{-1}\, dg \rangle = -(\Omega^{-1}\, dg)\cdot f .$$

By the definition of $[f,g]$ it is antisymmetric, so this is also $(\Omega^{-1}\, df)\cdot g$. By Theorem 4 of §4.3 again, $\theta_X dg = d(X \lrcorner dg) = d(dg(X)) = d(X\cdot g)$. Therefore, by (b) and (c),

$$[df,dg] = \theta_{\Omega^{-1}df}\, dg = d((\Omega^{-1}\, df)\cdot g) = d[f,g] . \qquad \text{QED.}$$

3. Symplectic structures in local coordinates

As we saw in paragraph 1, a coordinate Lie module must have an even number of coordinates to admit a symplectic structure.

<u>Theorem 3.</u> <u>Let</u> E <u>be a coordinate Lie module with coordinates</u> $q^1, p_1, \ldots, q^n, p_n$. <u>Then</u>

(1) $$\Omega = 2(dq^1 \wedge dp_1 + \dots + dq^n \wedge dp_n)$$

is a symplectic structure. Its components are given by the matrix

(2) $$\begin{pmatrix} 0 & -1 & & & \\ 1 & 0 & & & \\ & & \ddots & & \\ & & & 0 & -1 \\ & & & 1 & 0 \end{pmatrix}$$

If h *is a scalar then*

(3) $$\Omega^{-1} dh = \Sigma\left(\frac{\partial h}{\partial p_i}\frac{\partial}{\partial q^i} - \frac{\partial h}{\partial q^i}\frac{\partial}{\partial p_i}\right) .$$

We have

(4) $$\Omega \frac{\partial}{\partial q^i} = dp^i, \qquad \Omega \frac{\partial}{\partial p_i} = -dq^i$$

and

(5) $$\Omega^{-1} dq^i = -\frac{\partial}{\partial p_i}, \qquad \Omega^{-1} dp_i = \frac{\partial}{\partial q^i} .$$

If f *and* g *are scalars,*

(6) $$[f,g] = \Sigma\left(\frac{\partial f}{\partial p_i}\frac{\partial g}{\partial q^i} - \frac{\partial f}{\partial q^i}\frac{\partial g}{\partial p_i}\right) .$$

Proof. Everything else is a trivial consequence of (4), which is a trivial consequence of the definition of the wedge product. QED.

The theorem of Darboux says that on a manifold with a symplectic structure Ω one may always choose local coordinates so that Ω is given locally by (1). Thus all symplectic manifolds of a given dimension are locally the same. This is in strong contrast to the great variety of locally non-isometric Riemannian manifolds.

By Theorem 2(a), if h is any scalar on a symplectic manifold then the flow generated by the vector fields $\Omega^{-1}dh$ preserves the symplectic structure. Again this is in strong contrast to the Riemannian case, where there seldom exists a flow of isometries. Riemannian metrics are much more rigid than symplectic structures. Riemannian metrics admit a distinguished affine connection, and the group of automorphisms preserving an affine connection is always a Lie group (parameterized by finitely many variables). Symplectic structures do not admit a distinguished affine connection, and the local automorphisms preserving the symplectic structure form a pseudogroup (parameterized by a function).

4. Hamiltonian dynamics

Let M be a manifold, $T^*(M)$ its cotangent bundle. An element of $T^*(M)$ is simply a cotangent vector η_q at some point q of M, so that if $q^1,\ldots,q^n$ are local coordinates near q then

$$\eta_q = \Sigma\, p_i(q)(dq^i)(q) .$$

Now the q^i and p_i are functions on $T^*(M)$ and are in fact a local coordinate system, so that

$$\theta = \Sigma\, p_i dq^i$$

is a 1-form on $T^*(M)$. The 1-form θ is well-defined globally (it does not depend on the choice of local coordinates) since it has the invariant description

$$< \theta(\eta_q), X(\eta_q) > \; = \; < \eta_q, \pi_* X(\eta_q) >$$

where $X(\eta_q)$ is a tangent vector on $T^*(M)$ at the point η_q and $\pi\colon T^*(M) \longrightarrow M$ is the projection sending each η_q to q (and π_* is the induced mapping of the tangent bundles). Then $\Omega = -2d\theta$ is a symplectic structure on $T^*(M)$, and is given by (1) with respect to the local coordinates $q^1, p_1, \ldots, q^n, p_n$. Thus the cotangent bundle of an arbitrary manifold admits a natural symplectic structure.

In classical mechanics the configuration space of a mechanical system is a manifold M and $T^*(M)$ is the momentum phase space. If $q^1, \ldots, q^n$ are local coordinates on M then $p_1, \ldots, p_n$ are the conjugate momenta. The symplectic structure Ω knits together the coordinates and their conjugate momenta. In Hamiltonian mechanics one allows all transformations which preserve Ω even if the distinction between coordinates and momenta is lost.

The energy of a classical dynamical system is a scalar H on the momentum phase space. Hamilton's equations say that the time evolution of the dynamical system is given by the flow with generator $\Omega^{-1}dH$. By (3) this means that in local coordinates

$$\frac{dq^i}{dt} = \frac{\partial H}{\partial p_i}, \qquad \frac{dp_i}{dt} = -\frac{\partial H}{\partial q^i}.$$

Ralph Abraham gave a course at Princeton on this subject [8] last year and I will say no more. Note: my Ω is twice Abraham's ω and Sternberg's Ω is minus Abraham's ω. Therefore the different account of Poisson brackets vary slightly but all are such that one obtains the same formulas in local coordinates.

References

[8] R. Abraham and J. Marsden, Foundations of Mechanics, Benjamin (New York), 1967.

[9] Shlomo Sternberg, Lectures on Differential Geometry, Prentice-Hall (Englewood Cliffs, N.J.), 1964.

[10] G. W. Mackey, Mathematical Foundations of Quantum Mechanics, Benjamin (New York), 1963.

§9. Complex structures

1. Complexification

On an arbitrary differentiable manifold it makes sense to consider complex-valued scalars and tensors, covariant derivatives in the direction of a complex-valued vector field, etc. Algebraically, we do this as follows.

Let F^{oc} be the algebra obtained by joining an element i to F^o with $i^2 = -1$. (Thus F^{oc} is a field if and only if F^o contains no square root of -1.) If V is any F^o vector space we let V^c be the F^{oc} module obtained by extending coefficients to F^{oc}. Thus $V^c = V+iV$ and $i(X+iY) = -Y+iX$. We define complex conjugation on V^c by $\overline{X+iY} = X-iY$. If u is an F^o-multilinear mapping of $V_1 \times \ldots \times V_n$ into V it has a unique F^{oc}-multilinear extension, again denoted u, mapping $V_1^c \times \ldots \times V_n^c$ into V^c.

In particular, F^c is again a commutative algebra with unit over F^o; if E is an F module then E^c is an F^c module; if E is a Lie module so is E^c; if ∇ is an affine connection on E its extension ∇ is an affine connection on E^c.

2. Almost complex structures

Definition. An almost complex structure on an F module E is an F-linear transformation J of E into itself such that $J^2 = -1$.

If E is free of finite type with basis $X_1, \ldots, X_n$ and J is an almost complex structure then the components of J satisfy

$$\Sigma\, J_i^j J_j^k = -\delta_i^k$$

and conversely such J_i^j give an almost complex structure. If there is no scalar f in F such that $f^2 = -1$ then n must be even, since

$$(\det J)^2 = \det J^2 = \det(-1) = (-1)^n .$$

It must be emphasized that an almost complex structure J is very different from i. The transformation J maps E into itself whereas i maps E into E^c. The transformation J has a unique extension J to E^c as in the preceding paragraph, and $iJ = Ji$.

<u>Definition.</u> Let J be an almost complex structure on the F module E. We define $E_{1,0}$ to be the set of vector fields Z in E^c such that $JZ = iZ$ and $E_{0,1}$ to be the set of Z in E^c such that $JZ = -iZ$. Elements of $E_{1,0}$ are said to be of <u>type</u> $(1,0)$, those in $E_{0,1}$ of <u>type</u> $(0,1)$.

<u>Theorem 1.</u> <u>Let</u> J <u>be an almost complex structure on the</u> F <u>module</u> E. <u>Then</u> E^c <u>is the</u> F^c <u>module direct sum of</u> $E_{1,0}$ <u>and</u> $E_{0,1}$. <u>The projections onto</u> $E_{1,0}$ <u>and</u> $E_{0,1}$ <u>are given by</u>

$$P = \tfrac{1}{2}(1-iJ), \qquad \bar{P} = \tfrac{1}{2}(1+iJ) \tag{1}$$

<u>respectively. Complex conjugation is bijective from</u> $E_{1,0}$ <u>to</u> $E_{0,1}$.

<u>Proof.</u> It is clear that $P+\bar{P} = 1$, $P\bar{P} = \bar{P}P = 0$, and that $PZ = Z$ if and only if Z is of type $(1,0)$ and $\bar{P}Z = Z$ if and only if Z is of type $(0,1)$. The last statement in the theorem is obvious. QED.

Recall (§3.2) the notion of a derivation of (F,E).

<u>Theorem 2.</u> <u>Let</u> J <u>be an almost complex structure on the</u> F <u>module</u> E <u>and let</u> φ <u>be a derivation of</u> (F^c,E^c). <u>Then</u> $P\varphi\bar{P}$ <u>and</u> $\bar{P}\varphi P$ <u>are</u> F^c<u>-linear</u>.

<u>Proof.</u> $P\varphi\bar{P}fZ = P\varphi f\bar{P}Z = fP\varphi\bar{P}Z + P(\varphi f)\bar{P}Z = fP\varphi\bar{P}Z$, and similarly for $\bar{P}\varphi P$. QED.

3. Torsion of an almost complex structure

Definition. Let J be an almost complex structure on the Lie module E. The torsion T of J is defined by

$$(2) \qquad T(Z,W) = -\bar{P}[PZ,PW] - P[\bar{P}Z,\bar{P}W]$$

and the torsion tensor by $T(\omega,Z,W) = \langle\omega,T(Z,W)\rangle$. A complex structure on the Lie module E is an almost complex structure whose torsion is 0.

We may rewrite (2) as $T(Z,W) = P[PZ,PW] + \bar{P}[\bar{P}Z,\bar{P}W] - [Z,W]$.

Theorem 3. Let J be an almost complex structure on the Lie module E with torsion T. Then T is antisymmetric and F^c-bilinear. The module $E_{1,0}$ is a Lie module if and only if J is a complex structure.

For all X and Y in E^c,

$$(3) \qquad 4T(X,Y) = [JX,JY] - [X,Y] - J[JX,Y] - J[X,JY].$$

Proof. By Theorem 2, $\bar{P}[PZ,PW]$ and $P[\bar{P}Z,\bar{P}W]$ are F^c-bilinear, so that T is too (and consequently the torsion tensor is a tensor). T is clearly antisymmetric. If J is a complex structure then $\bar{P}[PZ,PW]$ and $P[\bar{P}Z,\bar{P}W]$ are each always 0, so $E_{1,0}$ and $E_{0,1}$ are Lie modules. If $E_{1,0}$ is a Lie module then so is $E_{0,1}$ since $[\bar{Z},\bar{W}] = \overline{[Z,W]}$, so that T is 0. The formula (3) is most easily proved by verifying it for X and Y in $E_{1,0}$ and for X and Y in $E_{0,1}$ and by observing that for X and Y of different types both sides of (3) give 0. QED.

Theorem 4. Let E be a coordinate Lie module with coordinates $x^1,\ldots,x^n$ and let J be an almost complex structure on E with components J_i^j. Then the components of the torsion tensor T are given by

$$T^k_{ij} = \frac{1}{2}\,\Sigma\{J^a_i(\frac{\partial}{\partial x^a}J^k_j) - J^a_j(\frac{\partial}{\partial x^a}J^k_i) + (\frac{\partial}{\partial x^j}J^a_i)J^k_a - (\frac{\partial}{\partial x^i}J^a_j)J^k_a\} .$$

Proof. The proof is trivial. QED.

4. Complex structures in local coordinates

Theorem 5. *Let* E *be a coordinate Lie module with coordinates* $x^1,y^1,\ldots,x^n,y^n$ *and let* J *in* E^1_1 *have components given by the matrix*

$$(4)\qquad \begin{pmatrix} 0 & -1 & & & \\ 1 & 0 & & & \\ & & \ddots & & \\ & & & 0 & -1 \\ & & & 1 & 0 \end{pmatrix}$$

Then J *is a complex structure. Let* $z^1 = x^1+iy^1,\ldots,z^n = x^n+iy^n$. *Then* $E_{1,0}$ *is a coordinate Lie module with coordinates* $z^1,\ldots,z^n$, *and the basis dual to* $dz^1,\ldots,dz^n$ *is given by*

$$(5)\qquad \frac{\partial}{\partial z^1} = \frac{1}{2}(\frac{\partial}{\partial x^1} - i\frac{\partial}{\partial y^1}),\ldots,\frac{\partial}{\partial z^n} = \frac{1}{2}(\frac{\partial}{\partial x^n} - i\frac{\partial}{\partial y^n}) .$$

Proof. It is clear that J is an almost complex structure. The elements (5) are simply P applied to $\partial/\partial x^1,\ldots,\partial/\partial x^n$, and it is clear that they are a basis of $E_{1,0}$. Since they commute, $E_{1,0}$ is a Lie module and J is a complex structure by Theorem 3. (This also follows from Theorem 4, since the components of J are constants.) QED.

The Newlander-Nirenberg theorem asserts that on a differentiable manifold with an almost complex structure J whose torsion is 0 (i.e., a complex structure as we have defined it) one can choose local coordinates in the neighborhood of any point so that J has the above form.

This theorem is the justification for calling an almost complex structure with torsion 0 a complex structure. In contrast to the Darboux theorem (§8.3), the Newlander-Nirenberg theorem is quite difficult and the result was unknown for a long time. Consequently terms such as pseudocomplex structure and integrable almost complex structure are used in many places for an almost complex structure with torsion 0. Once a manifold has a complex structure the theory of functions of several complex variables may be applied.

When using classical tensor notation when an almost complex structure is given, the convention is made that Greek covariant indices represent vector fields of type $(1,0)$, so that Greek covariant indices with a bar over them represent vector fields of type $(0,1)$. Instead of a bar, a dot or a star is sometimes used. Thus if T is the torsion tensor of the almost complex structure, $T^k_{\alpha\bar{\beta}} = 0$.

5. Almost complex connections

If J is an almost complex structure we let J be the tensor in E^1_1 given by $J(\omega,X) = \langle \omega, JX \rangle$. An affine connection ∇ such that $\nabla J = 0$ is called an <u>almost complex connection.</u> This is the same as requiring that $[\nabla_X, J] = 0$ or $[\nabla_X, P] = 0$ for all vector fields X.

<u>Theorem 6. Let</u> E <u>be a Lie module such that there exists an affine connection on</u> E, <u>and let</u> J <u>be an almost complex structure on</u> E. <u>If</u> ∇ <u>is an affine connection on</u> E <u>let</u>

$$\nabla^0_X = P\nabla_X P + \bar{P}\nabla_X \bar{P} .$$

<u>Then</u> ∇^0 <u>is an almost complex connection and</u> $\nabla = \nabla^0$ <u>if and only if</u> ∇ <u>is an almost complex connection. Let</u> T^0 <u>be its torsion and let</u>

$$\nabla^*_X = P(\nabla_X - T^\circ_X)P + \overline{P}(\nabla_X - T^\circ_X)\overline{P} .$$

Then ∇^* is an almost complex connection, and if ∇ is torsion-free then the torsion of ∇^* is the torsion T of the almost complex structure. J is a complex structure if and only if there is a torsion-free almost complex connection. If ∇ is any almost complex connection with torsion T' then the torsion T of the almost complex structure is given by

(6) $$T(Z,W) = \overline{P}T'(PZ,PW) + PT'(\overline{P}Z,\overline{P}W) .$$

Proof. $\nabla_X - \nabla^\circ_X = \overline{P}\nabla_X P + P\nabla_X\overline{P}$ and by Theorem 2 this is F^c-linear, so that ∇° is an affine connection. Therefore ∇^* is also an affine connection. They are almost complex connections since ∇°_X and ∇^*_X clearly commute with P. If $\nabla = \nabla^\circ$ then ∇ is an almost complex connection too, and conversely if ∇ is an almost complex connection then $\overline{P}\nabla_X P$ and $P\nabla_X\overline{P}$ are 0, so that $\nabla = \nabla^\circ$. Suppose that ∇ is torsion-free and let T^* be the torsion of ∇^*. For Z and W of type $(1,0)$,

$$T^\circ(Z,W) = P(\nabla_Z W - \nabla_W Z) - [Z,W] = -\overline{P}[Z,W] = T(Z,W) .$$

This term is annihilated by P, so for Z and W of type $(1,0)$ we have $T^*(Z,W) = T(Z,W)$. Similarly for Z and W of type $(0,1)$. Now suppose that Z and W are of different types, say $PZ = Z$ and $\overline{P}W = W$. Then

$$T^\circ(Z,W) = \overline{P}\nabla_Z W - P\nabla_W Z - [Z,W]$$

so that

$$T^*(Z,W) = \overline{P}\nabla_Z W - \overline{P}T^\circ(Z,W) - P\nabla_W Z + PT^\circ(W,Z) - [Z,W]$$

$$= T^\circ(Z,W) - T^\circ(Z,W) = 0 = T(Z,W) .$$

Therefore $T^* = T$ if ∇ is torsion-free. Since E has an affine connection, it has a torsion-free affine connection ∇ (§5.6) and so has an almost complex connection ∇^* whose torsion is T. Therefore if J is a complex structure there is a torsion-free almost complex connection. Conversely, if ∇ is a torsion-free almost complex connection then ∇_Z maps $E_{1,0}$ into itself (this is clearly true for any almost complex connection) so that if Z and W are in $E_{1,0}$ so is $[Z,W] = \nabla_Z W - \nabla_W Z$, and by Theorem 3, J is a complex structure. This also follows from the last statement of the theorem, which we now prove. For Z and W of type (1,0), the right hand side of (6) is

$$\overline{P}(\nabla_Z W - \nabla_W Z - [Z,W]) = -\overline{P}[Z,W] = T(Z,W) ,$$

and similarly for Z and W of type (0,1). If Z and W are of different types then both sides of (6) are 0. QED.

6. Kähler structures

We have discussed pseudo-Riemannian metrics, almost symplectic structures, and almost complex structures. We conclude our study of tensor analysis by discussing briefly the interrelationships among these three types of structure.

A pseudo-Riemannian metric is a bijective symmetric mapping $g: E \longrightarrow E'$, an almost symplectic structure is a bijective antisymmetric mapping $\Omega: E \longrightarrow E'$ and an almost complex structure is a bijective mapping $J: E \longrightarrow E$ such that $J^{-1} = -J$ (all mappings being F-linear). An almost Hermitian structure on a Lie module E is a pseudo-Riemannian metric g and an almost complex structure J such that $\Omega = g \circ J$ is antisymmetric; i.e. such that

$$(7) \qquad \Omega(X,Y) = g(JX,Y) = -g(X,JY) = -\Omega(Y,X)\,, \qquad X,Y \in E\,.$$

Then Ω is an almost symplectic structure, since Ω is clearly bijective with $\Omega^{-1} = J^{-1}\circ g^{-1}$. Since $J^2 = -1$, the relation (7) is equivalent to

$$(8) \qquad g(JX,JY) = g(X,Y)\,, \qquad X,Y \in E$$

or

$$(9) \qquad g(PZ,W) = g(Z,\bar{P}W)\,, \qquad Z,W \in E^c\,.$$

(For a manifold, the term almost Hermitian structure is usually reserved for the case that g is a Riemannian metric. Perhaps we should use the term almost pseudo-Hermitian structure, but we won't.) We may also give an almost Hermitian structure by means of a pseudo-Riemannian metric g and an almost symplectic structure Ω such that $J = g^{-1}\circ\Omega$ is an almost complex structure, or by an almost symplectic structure Ω and an almost complex structure J such that $g = \Omega\circ J^{-1} = -\Omega\circ J$ is symmetric and consequently a pseudo-Riemannian metric. We shall be democratic and indicate an almost Hermitian structure by (g,J,Ω) where g is pseudo-Riemannian, J is almost complex, Ω is almost symplectic, and $\Omega = g\circ J$. If (g,J,Ω) is an almost Hermitian structure and J is a complex structure then (g,J,Ω) is called a <u>Hermitian structure</u>. There is no name for an almost Hermitian structure in which Ω is a symplectic structure. An almost Hermitian structure (g,J,Ω) such that J is a complex structure and Ω is a symplectic structure is called a <u>Kähler structure</u>.

<u>Theorem 7.</u> <u>Let</u> (g,J,Ω) <u>be an almost Hermitian structure,</u> ∇ <u>the Riemannian connection,</u> T <u>the torsion of</u> J. <u>If</u> Z <u>and</u> W

in E^c *are of the same type then*

(10) $$g(Z,W) = 0 ,$$

(11) $$\Omega(Z,W) = 0 ,$$

(12) $$(\nabla_X \Omega)(Z,W) = 3d\Omega(Z,W,X) - 2\Omega(T(Z,W),X) .$$

If Z *and* W *are of opposite types then*

(13) $$(\nabla_X \Omega)(Z,W) = 0 .$$

Proof. Let Z and W be of type $(1,0)$. By (8),

$$g(Z,W) = g(JZ,JW) = g(iZ,iW) = -g(Z,W)$$

so that (10) holds in this case. Since J preserves types, (11) also holds in this case. Next observe that since ∇ is torsion-free,

$$\begin{aligned}(\nabla_Z J)(W) &- (\nabla_W J)(Z) \\ &= \nabla_Z(JW) - J\nabla_Z W - \nabla_W(JZ) + J\nabla_W Z \\ &= i[Z,W] - J[Z,W] = i(1+iJ)[Z,W] = -2J\bar{P}[Z,W] \\ &= 2JT(Z,W) .\end{aligned}$$

Since $\nabla g = 0$ and $\Omega = g \circ J$, this implies that

$$(\nabla_Z \Omega)(W) - (\nabla_W \Omega)(Z) = 2\Omega T(Z,W) .$$

That is,

$$(\nabla_Z \Omega)(W,X) - (\nabla_W \Omega)(Z,X) = 2\Omega(T(Z,W),X) .$$

Since ∇ is torsion-free, $d\Omega = \mathrm{Alt}\, \nabla\Omega$, so that the left hand side of this is $3d\Omega(Z,W,X) - (\nabla_X \Omega)(Z,W)$, so that (12) holds for Z and W of type $(1,0)$. By taking complex conjugates we see that (10), (11), and (12) also hold for Z and W of type $(0,1)$.

Now let Z and W be of opposite types, say $PZ = Z$ and $\bar{P}W = W$. Then

$$(14)\qquad \begin{aligned}(\nabla_X\Omega)(Z,W) &= X\cdot\Omega(Z,W) - \Omega(\nabla_X Z,W) - \Omega(Z,\nabla_X W)\\ &= X\cdot\Omega(Z,W) - \Omega(\nabla^o_X Z,W) - \Omega(Z,\nabla^o_X W)\\ &= (\nabla^o_X\Omega)(Z,W)\ ,\end{aligned}$$

where ∇^o is as in Theorem 6 and we have used (9). By the definition of ∇^o it is clear that each ∇^o_X commutes with J, so that $\nabla^o J = 0$. Now ∇^o_X differs from ∇_X by $P\nabla_X\bar{P} + \bar{P}\nabla_X P$. Since $\nabla g = 0$, to show that $\nabla^o g = 0$ we need only show that $P\nabla_X\bar{P}$ and $\bar{P}\nabla_X P$ (which are F^c-linear) are g-antisymmetric. But, by (9) and (10),

$$\begin{aligned}g(P\nabla_X\bar{P}U,V) &= g(\nabla_X\bar{P}U,\bar{P}V)\\ &= X\cdot g(\bar{P}U,\bar{P}V) - g(\bar{P}U,\nabla_X\bar{P}V) = -g(U,P\nabla_X\bar{P}V)\end{aligned}$$

and similarly for $\bar{P}\nabla_X P$. Thus $\nabla^o g = 0$ and so $\nabla^o\Omega = \nabla^o(g\circ J) = 0$. By (14), $(\nabla_X\Omega)(Z,W) = 0$ for $PZ = Z$ and $\bar{P}W = W$, and similarly for $\bar{P}Z = Z$ and $PW = W$. QED.

We remark that the last part of the proof shows that if (g,J,Ω) is an almost Hermitian structure then there is an affine connection ∇^o such that $\nabla^o g = 0$, $\nabla^o J = 0$, and $\nabla^o\Omega = 0$.

<u>Theorem 8.</u> <u>Let</u> (g,J,Ω) <u>be an almost Hermitian structure,</u> ∇ <u>the Riemannian connection. If</u> $\nabla J = 0$ <u>or</u> $\nabla\Omega = 0$ <u>then</u> (g,J,Ω) <u>is a Kähler structure. Conversely, if</u> (g,J,Ω) <u>is a Kähler structure,</u> ∇ <u>the Riemannian connection, then</u> $\nabla J = 0$ <u>and</u> $\nabla\Omega = 0$.

<u>Proof.</u> Since $\Omega = g\circ J$ and $\nabla g = 0$, if $\nabla J = 0$ or $\nabla\Omega = 0$ then $\nabla J = 0$ and $\nabla\Omega = 0$. The Riemannian connection is torsion-free, so J is a complex structure by Theorem 6 and Ω is a symplectic

structure by §8.1. Therefore (g,J,Ω) is a Kähler structure if $\nabla J = 0$ or $\nabla\Omega = 0$.

Conversely, let (g,J,Ω) be a Kähler structure, so that the torsion T of J is 0 and $d\Omega = 0$. By Theorem 7, $\nabla\Omega = 0$ and so $\nabla J = 0$ also. QED.

Complex projective space has a Kähler metric. Complex projective algebraic varieties without singularities are complex analytic submanifolds of complex projective space and so have an induced Kähler metric. Hodge theory was developed primarily for this situation. Since $\nabla\Omega = 0$ the operators L and Λ given by $L\alpha = \Omega\wedge\alpha$ and $\Lambda = L^*$ commute with the Laplace-deRham operator. By the theorems of Hodge and deRham, L and Λ act on the real cohomology ring, and place strong restrictions on the real cohomology of a non-singular complex projective algebraic variety (see [7,§7], [11], [12]). Also, if we define C on A^r by

$$(C\alpha)(X_1,\ldots,X_r) = \alpha(JX_1,\ldots,JX_r)$$

then $C^2 = (-1)^r$ on A^r and C commutes with Δ since $\nabla J = 0$. It follows that odd-dimensional Betti numbers of compact Kähler manifolds are even.

References

[11] André Lichnerowicz, Théorie globale des connexions et des groupes d'holonomie, Consiglio Nazionale delle Ricerche, Monografie Matematiche 2, Edizioni Cremonese, Rome, 1955.

[12] André Weil, Introduction à l'étude des varietés kaehlériennes, Hermann, Paris, 1958.

Ingram Content Group UK Ltd.
Milton Keynes UK
UKHW022000300623
424377UK00008B/784